Lessons from the

Sand

Lessons from the
Sand

Charles O. Pilkey & Orrin H. Pilkey

The University of North Carolina Press   CHAPEL HILL

FAMILY-FRIENDLY SCIENCE ACTIVITIES
YOU CAN DO ON A CAROLINA BEACH

A Southern Gateways Guide

Cover illustration by Charles O. Pilkey

Library of Congress Cataloging-in-Publication Data
Names: Pilkey, Charles O. | Pilkey, Orrin H., 1934–
Title: Lessons from the sand : family-friendly science activities you can do on a Carolina beach /
Charles O. Pilkey and Orrin H. Pilkey.
Other titles: Southern gateways guide.
Description: Chapel Hill : The University of North Carolina Press, [2016] | Series: A Southern gateways
guide | Includes bibliographical references and index.
Identifiers: LCCN 2015040682 | ISBN 9781469627373 (pbk : alk. paper) | ISBN 9781469627380 (ebook)
Subjects: LCSH: Beaches—Study and teaching—North Carolina. | Beaches—Study and teaching—
South Carolina. | Science—Experiments.
Classification: LCC GB459.4 .P55 2016 | DDC 552/.5—dc23 LC record available at
http://lccn.loc.gov/2015040682

In memory of Sharlene Pilkey,

1936–2015

Contents

Foreword viii

Acknowledgments xi

INTRODUCTION 1

HOW TO USE THIS BOOK 3
What You Need 4
Beach Safety 5
The Invisible Pollution Problem 6

1 WAVES 7
1 Wave Height 8
2 The Wind and the Waves 12
3 Surf 17
4 Longshore Currents 22
5 Rip Currents 27
6 Tides 30

2 MOVING BEACHES 37
7 Sea Breeze 38
8 Dunes 43
9 Ancient Shorelines 47
10 Beach Erosion 51
11 Barrier Islands 56

3 SAND 61
12 Sand 62
13 Black Sand 66
14 Barking Sand 70
15 Soft Sand 72

4 BEACH FEATURES 75
16 Ripple Marks 76
17 Volcanoes on the Beach 80
18 Beach Tracker 85
19 The Layer Cake Beach 91
20 Groundwater 95

5 SHELLS 99
21 Shells 100
22 Black Shells 110
23 Brown Shells 112
24 Shell Roundness 115
25 Shell Orientation 118
26 Fossils 121

6 LIFE AT THE BEACH 127
  27 Life in the Fast Lane 128
  28 Birds at the Beach 132
  29 Plants and Salt 143
  30 Murder Mystery 155

7 THE ENVIRONMENT 159
  31 Replenished Beaches 160
  32 Concrete Beaches 164
  33 Litter 168
  34 Beach Driving 172
  35 Ocean Acidification 174

8 RAINY DAY ACTIVITIES 177
  36 Plankton 178
  37 Saltwater 183
  38 Museum Visit 187
  39 Google Earth 192

9 SEEING THE UNSEEN 197
  40 Night 198
  41 Do Your Own Activity 204

Glossary 205
Geologic Time Scale 216
The Edge of the Wild 217
    Beach Conservation 217
    Books 219
Photo Credits 220
About the Authors 221

Foreword

The Child is father of the Man.
—William Wordsworth

The health of the sea and human, physical and spiritual well-being are closely linked. The sea provides us with food, medicine, oxygen, climate control and a surface for transporting goods. The sea is also a place of great beauty and mystery, especially where it meets the land at a beach.

Children, born with a sense of wonder, instinctively sense their intimate connection to nature. They revel in the miracle of sand and sea. Conversely, we adults are so absorbed in our engineered and technological world that we have allowed ourselves to drift away from this spontaneous, joyful embrace of nature. We have forgotten the innate wisdom of childhood. Nevertheless, by sustaining a dialogue with our children, we can yet revive our dormant curiosity, rekindle our sense of wonder and reconnect ourselves to the natural world.

Lessons from the Sand, written by a father and son, is a book as esthetically pleasing as it is rooted in good science. We believe a book like this can be a pathway to many fruitful conversations and exchanges between parents and children. Combining humor and beautiful illustrations with hands-on activities, this book explains the science of a Carolina beach in a manner that is lucid, fun and informative for readers of all ages. Education is the key to protecting our beaches, as the more we understand something, the more we appreciate it and are moved to protect it.

The Santa Aguila Foundation is proud to have made *Lessons from the Sand* possible. The foundation is a U.S. nonprofit organization dedicated to the preservation of shorelines around the world. Through information and education, we attempt to raise greater awareness and understanding of our coastal environment.

The very heart and essence of our charitable organization's mission is youth. We strongly believe that it is crucial for children to learn about the importance of shorelines, to be educated about the science of beaches and to be empowered to protect our coastal environment.

Our education efforts include management of the beach website coastalcare.org; the publication of three other books by Orrin Pilkey— *The World's Beaches, Global Climate Change: A Primer* and *The Last Beach*; and support for the award-winning documentary film *Sand Wars*.

We hope you will enjoy this book and take as much pride as we do in defending and protecting our coastal environment.

Please visit www.coastalcare.org for further information.

THE SANTA AGUILA FOUNDATION

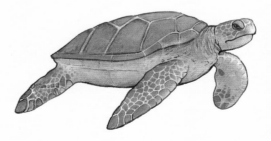

Acknowledgments

We are deeply grateful to the Santa Aguila Foundation for their support of the publication of this book. We would also like to express our appreciation for all that the foundation is doing to protect and preserve the world's beaches for future generations. Norma Longo helped us gather data on some aspects of Carolina beaches. She also helped with some of the editing. We are grateful to Bill Neal, Joe Kelley, Andrew Cooper and Miles Hayes who helped to bring this book to fruition by contributing photographs and ideas. Our children and grandchildren willingly acted as test subjects for some of the activities as we trekked down the Carolina coast. We thank our mates, Yuko and Sharlene, for their patience and support and their encouragement of our efforts in the long road to completion of this book. Sharlene passed away as we were finishing the writing, which is why this book is dedicated to her.

CHARLES PILKEY
ORRIN PILKEY

Explore. . . . Discover. . . . Wonder!

Lessons from the
Sand

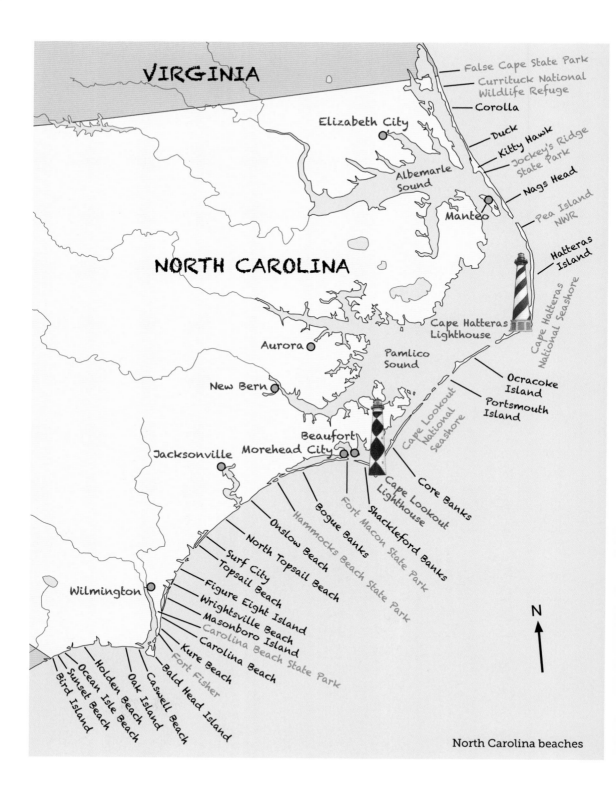

North Carolina beaches

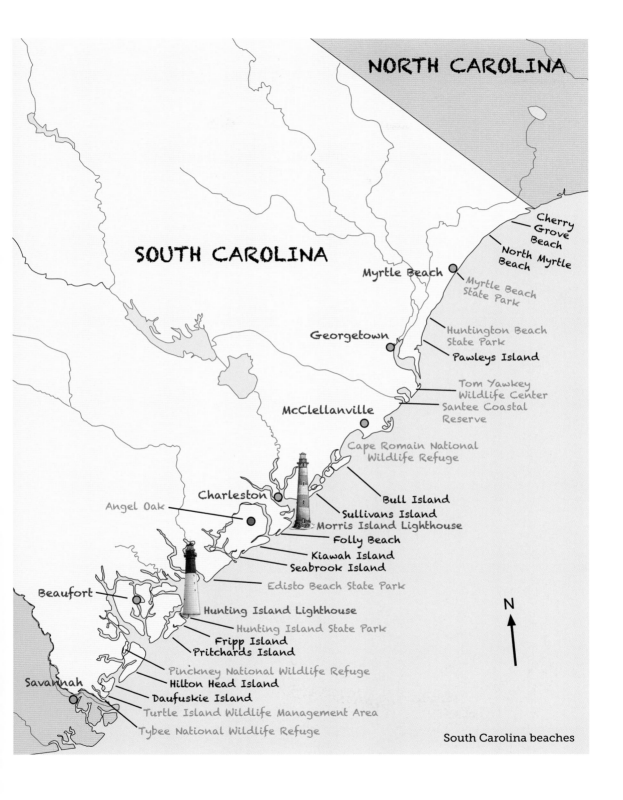

NORTH CAROLINA

SOUTH CAROLINA

Cherry Grove Beach

North Myrtle Beach

Myrtle Beach

Myrtle Beach State Park

Georgetown

Huntington Beach State Park

Pawleys Island

Tom Yawkey Wildlife Center

Santee Coastal Reserve

McClellanville

Cape Romain National Wildlife Refuge

Bull Island

Sullivans Island

Morris Island Lighthouse

Folly Beach

Kiawah Island

Seabrook Island

Charleston

Angel Oak

Edisto Beach State Park

Beaufort

Hunting Island Lighthouse

Hunting Island State Park

Fripp Island

Pritchards Island

Pinckney National Wildlife Refuge

Hilton Head Island

Savannah

Daufuskie Island

Turtle Island Wildlife Management Area

Tybee National Wildlife Refuge

N

South Carolina beaches

Introduction

I was six when our family moved to Sapelo, a wildly beautiful island shared equally by University of Georgia researchers, a community of Gullah-speaking African Americans and a nonhuman community of bears, deer, raccoons, alligators and diamondbacks so big they looked as if they could swallow a kid whole. From the shadows of the old forest rose ancient Indian mounds, a ruined slave plantation and the white-stucco luxury of the Reynolds Mansion. Rumors abounded of buried treasure and the ghosts of long-dead conquistadors. For reasons none could fathom, whales stranded themselves on the beach, sea turtles crawled out of the watery depths of midnight to lay eggs and hurricanes blew in from unguessed horizons, depositing layers of mud and decaying crabs on living room floors. For a young mind just getting acquainted with the world, Sapelo was a place of wonder and mystery, an Edenic wilderness granting a uniquely direct experience of nature—the kind of experience that might inspire a child to become a scientist or an artist, the kind of experience becoming rare in an increasingly digitalized world.

I remember only a little of the day that has long since passed into family lore. We were walking on the Sapelo beach looking for shells. My mother picked up a black shell and asked a simple question, "Why are some shells black?" The question puzzled my father. Apparently, no one had thought to ask the question before. I imagine he gave a vague reply about chemical reactions with iron and the like. But the question remained, nagging at the back of his thinking like an itch that couldn't be reached, until some years later its full import finally became clear.

Shells turn black in an oxygen-poor environment. In the beach ecosystem, such an environment occurs only in a salt marsh. Therefore, finding black shells on an undisturbed beach means that the beach has moved over a preexisting salt marsh. This was the first hint that barrier islands might be migrating and became the opening salvo in the ongoing battle between those wishing to develop barrier islands and those, recognizing the inherent instability and beauty of the beach ecosystem, wishing to keep the islands natural.

In such a fashion does science advance; someone with curiosity enough asks a question, and someone with energy enough finds the answer. Why do apples fall from trees? Why are finches on adjacent islands in the Galapagos similar yet different? Why does the west coast of Africa appear to fit so snugly with the east coast of South America? Behind every scientific advancement stands one person who asked a question none had yet thought to ask.

Although some of the experiments in this book require identification of plants, animals and minerals, the book is not a field guide. It is an activity book with an emphasis on learning by doing. We encourage families with young children (from kindergarten to middle school and beyond) to learn about Carolina beaches through observation and experiment, by direct physical contact with the beach ecosystem and above all, by asking questions and seeking the answers. Nature is too beautiful, varied and mysterious to be bound by memorized facts dredged from a textbook. Read a child a list of minerals found in beach sand and watch the lights go out. Curiosity curls up in a ball and goes to sleep. Bring that child to the beach to look for sharks' teeth and curiosity awakens, purrs contentedly and sets forth to explore the world. Watch the junior scientist stir to life.

Not every child can grow up at the edge of the wild in a place like Sapelo Island that nurtures a young human's inborn interest in nature. Most of us live in urban settings, and the untamed landscapes of the Southeast have largely vanished, melted away like snow under the unrelenting heat of development. But even a developed beach offers a view that is half wild. And it doesn't take much to fire the imagination of a young mind. All that is needed are a few simple tools, a few grains of sand, an ocean, a sky and curiosity as limitless as the sea. Explore. . . . Discover. . . . Wonder . . .

CHARLES PILKEY

How to Use This Book

The focus of this book is the beaches of North and South Carolina, yet many activities apply to sandy beaches anywhere in the world. The activities are accompanied by a series of questions. Some of the questions can be answered by completing the activities. Some questions are thought experiments that require thinking but no experimentation. And in some cases we don't know the answers to the questions, or the answers may vary from beach to beach.

One of the first problems you might run into is that many of the beaches in the Carolinas are replenished (or artificial). The sand was pumped or trucked in to replace an eroding beach. The shells and the critters living within the sand will not be the same as those living on the original natural beach. It's still possible to do many of the activities on artificial beaches, but natural ones are preferable. State and national parks are a good bet, the largest of which are the Cape Lookout and Cape Hatteras National

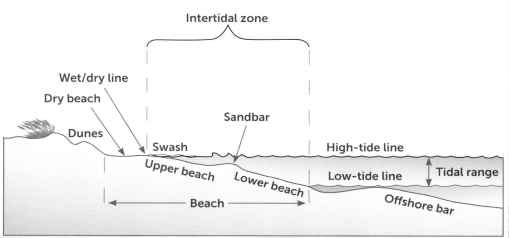

Profile of a Carolina beach.

Seashores in North Carolina and the Cape Romain National Wildlife Refuge in South Carolina.

Don't feel like you have to do the activities in any particular order. Thumb through the book and pick the ones you are most interested in. Take advantage of the glossary, where key terms highlighted in the text are explained in greater detail. Using a little imagination you may be able to design a better way to answer the questions that we pose or pursue an activity more deeply. Many of the activities would make excellent school science projects.

Always respect nature on the beach; don't feed birds, don't collect living shells, don't disrupt birds that are feeding or nesting, don't trample the dune vegetation, don't litter and clean up the litter already there. Collect only a few good shells. Leave the rest for nature and for others to enjoy.

What You Need

We recommend that you maintain a sketchbook journal (also called a nature journal) that includes drawings and notes. Such a journal could be useful for memorizing beach features and wildlife, especially if the journal is maintained for several seasons. Another approach would be to keep a digital journal with photographs, writings and videos. The following is a list of things to bring to the beach for the activities:

- Small shovel or trowel
- Tape measure
- Notebook/sketchbook and pencils
- Compass
- Binoculars (optional)
- Magnifying glass
- Microscope (available at bookstores, toy stores or online for less than $30)
- Small plastic sample bags

- Pocketknife
- Magnet
- One or two oranges
- Hydrometer (available at pet stores and some tackle shops)
- A jar of vinegar
- A watch or timer that measures seconds
- Kitchen scale (optional)
- Field guides for birds, beach plants, shells, etc. (optional)

Beach Safety

WHEN IN DOUBT, DON'T GO OUT. Are you really capable of handling the waves today? Know your limits. It's best to swim where a lifeguard is nearby. Don't go in the water or on the beach during a storm, especially if lightning is present.

INEXPERIENCED SWIMMER? Use a life preserver.

USE THE BUDDY SYSTEM. Pair up with a buddy and keep track of one another. Make sure an adult swimmer is nearby and watching.

WATCH OUT FOR STRONG CURRENTS AND KNOW HOW TO ESCAPE THEM. (See Activities 4, 5 and 6).

BE CAREFUL OF JELLYFISH (SEA JELLIES). Most (but not all) jellyfish stings can be treated by rinsing with seawater (not fresh water), vinegar or baking soda paste. Use a credit card to scrape the stinging cells off the skin. Seek medical help for any severe allergic reaction (difficulty breathing, vomiting, headache).

STINGRAYS AND SHARKS. Avoid stepping on stingrays by shuffling your feet to flush them out of the sand. Reduce chance encounters with sharks by 1) never swimming near a pier and 2) keeping a watchful eye for fins breaking the surface.

The Invisible Pollution Problem

Many people are unaware of the variety of bacteria, viruses, parasites and other microorganisms that inhabit beach sand and may cause illness. Where do these microorganisms come from? They come from many sources: bird droppings, dogs whose owners allow them to poop on the beach, people who defecate in the water, contaminated water delivered to the beach by storm drains and bacteria and other disease-causing organisms that naturally live in sand and saltwater. To minimize illness (stomachaches, coughs and sometimes more serious problems) consider the following recommendations from scientists:

- Don't swim with cuts or open sores.
- Don't swim if you are ill.
- Avoid long contact with beach sand by sitting on beach towels or chairs, and never, ever getting buried in sand.
- Avoid swimming after a heavy rain.
- Don't swim near drainage pipes or drainage streams.
- Wash your hands before eating.

1

Waves

I see the waves upon the shore,
Like light dissolved in star-showers thrown.
—Percy Bysshe Shelley
(nineteenth-century British poet
who drowned in a storm at sea)

Nothing conveys the mystery and power of the sea more than waves. For children playing in the surf they are liquid magic. For the poet their sound inspires metaphor. For the surfer they are endless delight. But to a geologist waves are more than a source of poetic wonder and recreational pleasure. They are the energy for the currents that push sand up and down the shoreline and the prime mover that creates, maintains and sometimes even destroys beaches. Without waves, beaches as we know them wouldn't exist.

The activities in this chapter talk about different kinds of waves, how they are formed and why they are important for sand transport, beach erosion and the creation of longshore currents. The chapter also introduces the scientific method and invites you to create a hypothesis to explain the sounds of breaking waves. The last activity is about tides, which are in effect giant, planetwide waves that whirl around the world in a cosmic choreography with the sun and the moon and the spinning earth.

Wave Height

Water is the driving force of all nature.
—Leonardo da Vinci
(the artist who painted the *Mona Lisa*)

Most waves on the sea are made by wind. As the wind blows over the surface of the sea it drags and pushes against the water, forming waves that move in the same direction as the wind. **Wave height** is determined by 1) wind strength, 2) wind duration and 3) the distance across the sea that a wind can blow (known as **fetch**). When the wind blows harder, for a longer period of time or has a greater fetch, both wave height and **wavelength** increase. Wave height is defined as the vertical distance from **wave trough** to **wave crest**. Wavelength is the distance from crest to crest or from trough to trough.

Wave height is notoriously difficult to accurately estimate. Surfers tend to understate wave height, while swimmers often overstate it. Is there a simple way to calculate wave height? Aside from surfers, why should beachgoers and coastal residents be concerned about the height of waves?

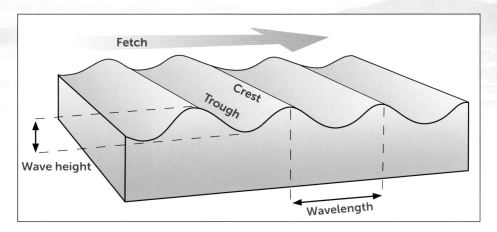

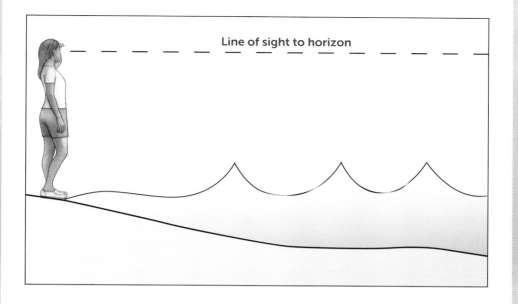

Line of sight to horizon

Estimating Wave Height: The Horizon Method

WHAT YOU NEED Patience and a good pair of eyes.

WHAT TO DO Stand on the beach as close to the water as you can get and look straight to the horizon. Are the crests of the waves above or below the horizon?

- If the crests are below the horizon then the height of the waves is less than your body height.
- If the crests are above the horizon then the waves are higher than your height.
- If the crests are level with the horizon the waves are more or less the same as your height.

The Surfer Method

WHAT YOU NEED A beach with surfers.

WHAT TO DO Assume all surfers stand six feet tall. Position yourself at the water's edge. Wait until a surfer is in a trough. If the wave crest is level with the standing surfer's head, it's about a six-foot wave. If the crest is over the surfer's head, the wave is higher than six feet. If the crest is twice the height of a surfer, it's a twelve-foot wave, and so on. You can also compare wave heights to people standing on a pier or to the pier itself.

6 feet

This wave is around 6 feet high.

WHY WAVE HEIGHT MATTERS

Knowing how to determine local wave height is important if your family is planning to go swimming, especially if young children are present. Bigger waves can create powerful currents and dangerous undertows. Wave height also affects the amount of sand moved about, location and shape of the beach and grain size. That's because bigger waves have more energy. They can move more sand around and can move larger sand grains than smaller waves. And of course, big waves sometimes cause big erosion problems (see Activity 10).

BIG WAVES

On average, waves are higher at Cape Hatteras and get smaller as you go south toward Hilton Head beach. But any beach may have higher than normal waves if a storm is nearby. And occasionally a beach may present unfamiliar and infrequent wave phenomena such as the following:

Tsunamis (soo-*na*-mees) are large waves caused by earthquakes, landslides or meteor strikes. They sometimes occur in New England but are extremely rare in the Carolinas. In 1929 an underwater landslide off eastern Canada triggered a 20-foot tsunami in Newfoundland. Fortunately, by the time the waves reached Charleston, South Carolina, they were only a few inches high.

Meteotsunamis (mee-tee-oh-soo-*na*-mees) are waves that resemble tsunamis but are caused by weather disturbances. Scientists are still trying to understand how they form. On June 13, 2013, a six-foot wave, believed to be a meteotsunami, swept people off rocks at Barnegat Inlet, New Jersey. The wave was recorded by tide gauges as far south as North Carolina.

Rogue waves happen when two waves briefly merge to create a larger wave. In deep water they can exceed 100 feet in height and are believed to have sunk countless ships that mysteriously vanished at sea. Rogue waves on the beach are often only slightly higher than the surf on a particular day and so go unnoticed. But occasionally they are far higher than expected. In 2010 a rogue wave washed over a seawall at a Mavericks surf contest in California, injuring several spectators.

EXPLORE YOUR WORLD

Learn more about tsunamis, meteotsunamis and rogue waves by searching online for news about large waves that suddenly appear on the world's beaches. Check YouTube for interesting videos about unusually big waves.

ACTIVITY 2

The Wind and the Waves

Trying to understand is like straining through muddy water.
Have the patience to wait! Be still and allow the mud to settle.
—Lao Tzu (Chinese philosopher and poet)

Since waves are formed by winds zooming over the sea and since winds can blow in any direction, waves can come from many directions. Sometimes waves move far from where the wind first created them. This is why waves on the beach may be traveling in a different direction than local winds.

Wave direction is the compass direction from which a wave is coming. Wave period is the time it takes for two consecutive wave crests to pass a point (such as a pier piling). When waves are first generated by wind, they are chaotic and choppy. These kinds of waves are referred to as sea and usually have periods of three to eight seconds. Eventually waves leave the area where they were generated and become better organized into long, regular, symmetrical wave trains. These are called swell and have periods ranging from 10 to 30 seconds. Swells form in the deep ocean far from land and are capable of moving across an ocean for hundreds, even thousands of miles.

Before there were satellite-based weather forecasts, old-timers studied the waves for signs of an approaching storm. They noted wave height. They noted wave direction, and they timed the period of the waves. By studying waves they knew when a storm was at hand and could anticipate its intensity. How was that possible?

Wave Direction

WHAT YOU NEED A notebook, a pencil, a good compass and a pier.

WHAT TO DO Stand on a pier beyond the surf. Use a compass to determine the direction the waves are coming from. First place your compass flat on your palm directly in front of your body and face the incoming waves. Rotate the compass until the needle is over the "N" or "North" mark. In your mind draw an imaginary line from the incoming waves to the center of the compass. The angle that line makes with the north-facing compass needle is the wave direction. In the illustration below the wave direction is 60 degrees east of north or 60 degrees northeast.

Record your readings in a notebook. Is wave direction the same as wind direction? If yes, then local wind may be generating the waves. If no, then the waves must have originated from another source, perhaps from a storm far out at sea.

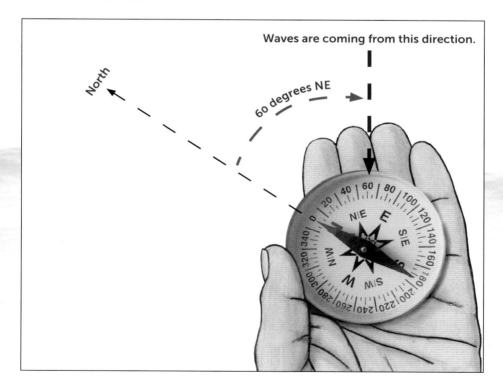

Wave Period—Timing Is Everything

WHAT YOU NEED A watch or digital timer with a second hand, a notebook and a pier.

WHAT TO DO Now that you've determined the direction of the wind and waves, measure the wave period. Stand near the end of a pier. Record the time it takes for two consecutive wave crests to pass your location. Take 10 measurements and calculate the average. Are the wave conditions sea or swell? Remember a "sea" is generated by local winds and is choppy and confused. Swells are long-period waves coming from a distant wind source, sometimes hundreds of miles away. They are well defined and have a smoother appearance than sea.

Do this same exercise at high tide and low tide and on different days and record your data. If you are lucky and a small storm comes by, measure the wave period before and after the storm (be mindful of lightning). Watch the Weather Channel and see if any storms are brewing offshore. If so, you might see some good swells. Hurricanes often swing to the east and miss the Carolinas, leaving behind swells for surfers (and sighs of relief from coastal residents).

Date	Time	Wave height	Wave period	Wave direction
7/12/15	10 a.m.	3–4 ft.	8 sec.	110° east

So how could old-timers predict an approaching storm without the benefit of modern technologies? First of all, they knew from long experience that bigger waves (like storm waves) have longer wave periods. They also knew that larger than normal waves often precede storms. And finally, they paid attention to the **atmospheric pressure**. If the barometer (an instrument that measures atmospheric pressure) continued dropping while the skies darkened, it was time to batten down the hatches and head for higher ground.

Bending Waves

Most waves approach a shoreline at an angle. The part of a wave that first reaches shallow water begins to slow down because of friction with the sea bottom. The part of the wave still in deep water doesn't slow down. The result is that the wave bends or refracts until it becomes almost parallel to the shoreline. This is known as wave refraction and is best seen from the end of a pier or an upper floor of a seaside hotel.

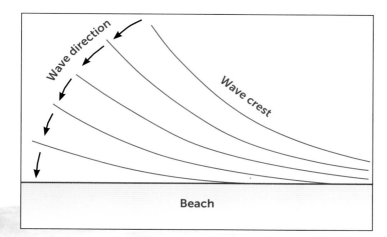

Wave refraction is the bending of waves as they approach the shore.

Bigger waves have longer wave periods. Tsunamis can have periods ranging from minutes to hours.

Waves Sure Are Swell

If your view of the sea is a good one, you may see more than one set of waves coming toward the shore. One is likely to be local waves. The other, formed by winds somewhere else, is likely to be swell. Look closely and you may even spot a third set of waves. Occasionally the waves you see are not made by the wind but are from the wake of a ship, passing just over the horizon.

Two sets of waves seen from the pier at Myrtle Beach State Park. One set is approaching from the southeast. The other set is coming from the south.

DID YOU KNOW?

Wave direction is not just a useful indicator of the general location of a possible storm. Wave direction is also critical to the formation of both **longshore currents** and **rip currents**. Visit Activities 4 and 5 to learn why.

ACTIVITY 3

Surf

The three great elemental sounds in nature are the sound of rain,
the sound of wind . . . and the sound of outer ocean on a beach.
—Henry Beston (American nature writer)

As waves approach the shore they drag on the sea floor and slow down. For this reason, their **wavelength** decreases and their height increases. Eventually, the waves get so steep they become unstable. At that point, they break under their own weight and become what we call surf. Oceanographers recognize three types of breaking waves:

Spilling breakers form when the upper part of the **wave crest** "spills" down the face of the wave. These are the most common type of waves and form on relatively flat beaches.

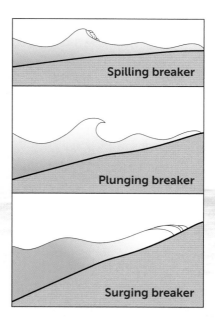

The **plunging breaker** is the classical wave loved by all surfers and the subject of countless movies and photographs. These are the most dangerous waves to swimmers and tend to occur on beaches that are somewhat steep.

Surging breakers are waves that never break. Instead, they rush up a beach as foam, slowly losing momentum before sliding back to sea. Surging breakers occur on beaches with very steep slopes.

Sandbars

Frequently on Carolina beaches there will be a line of white water offshore formed by waves breaking on sandbars. Two or even three parallel sets of bars can sometimes be seen. Occasionally a wave will partially break over a sandbar as a spilling breaker and still have enough energy to continue to the shore and break again as a plunging breaker.

Waves can be beautiful, scary, hypnotic and fun (especially if you like to surf). They can also be quite complex, and the science of waves is still being worked out. To explain natural events (like breaking waves) scientists use the scientific method, which, along with observation skills, curiosity and imagination (plus a healthy dose of skepticism), is the basic tool needed to unravel nature's mysteries.

Waves breaking on parallel sets of sandbars near Avon, N.C.

Spilling breaker

Plunging breaker

Surging breaker

Studying the Surf

WHAT YOU NEED A notebook, a beach with surf and lots of patience.

WHAT TO DO Observe the surf over a period of hours and days. If possible visit a beach a few days before or after a major storm and note any changes in wave type. What types of breaking waves do you see? What do they tell you about the beach slope? Is the surf different at high versus low tide? Does wind strength or wind direction affect the surf? Which breaking wave do you think causes the most coastal erosion?

The Scientific Method

Scientists use the scientific method to understand how nature works. A simplified version of the method is as follows:

1 Ask a question.
2 Make a hypothesis (an educated guess) that seems to answer your question.
3 Test your hypothesis with an experiment.
4 Decide if the hypothesis is correct or not.

For example, you ask, "Why is some sand red?" Your hypothesis is that red sand consists of red minerals. You test your hypothesis by looking at sand under a magnifying glass. Because you can clearly see red mineral grains you conclude your hypothesis is likely correct. See the glossary to learn more about the scientific method.

Sometimes experiments aren't possible. For example, when Charles Darwin developed his theory of evolution, it was impossible for him to physically observe plants and animals changing over the course of millions of years. Instead of experiment, Darwin used careful observation and deep thinking (along with the fossil record) to support his ideas.

What Makes the Sound of Waves?

WHAT YOU NEED A notebook, sharp ears and deep-thinking skills.

WHAT TO DO Stand at the edge of the sea, close your eyes and listen to the sound(s) of waves. List the different water sounds you hear. Make a hypothesis to explain how waves make sounds.

When making a hypothesis, scientists sometimes find it useful to break a natural system down to its basic components. For example, the basic parts of the system we call the surf are water, air, sand and the marine life that lives in the waves. One or more of these components somehow create the sound of waves. The question is which one(s) and how?

I Can Hear the Sea

Can you really hear the sound of the sea when you hold a shell next to your ear? Some have claimed the sound is really your own blood pumping through your veins. Others believe that sound waves from surrounding noise are funneled by the shell, creating the illusion of surf sound. What's *your* hypothesis? Can you think of an experiment to prove or disprove your hypothesis?

More about Surf

It's possible to see different types of breaking waves at the same beach during different times in the tide cycle. A strong onshore wind (blowing from sea to land) for several hours can create higher than normal tides, causing waves to break higher up on the beach. A strong offshore wind (from land to sea) would have the opposite effect. As any surfer will tell you, the best riding waves are early in the morning. That's because the morning wind is usually weak and blows offshore (toward the sea). In the afternoon the sea breeze (see Activity 7) blowing toward land makes the waves choppy and difficult to catch.

In theory, surging breakers should cause the most beach erosion because they slam into a beach without slowing down. In reality, most beach erosion comes from storm waves, which are often chaotic and defy easy classification. A spilling wave spends more time and energy dragging across the sea floor and should therefore deposit more sand on a beach.

Listen closely to the surf and you should hear three distinct sounds:

1 the boom of a breaking wave
2 the sound of foamy, white water heading toward the shore
3 the hissing of a wave as it recedes from a beach.

Scientists believe the sound of a breaking wave comes from air being compressed as the wave collapses, while the sound of foamy water is caused by thousands of air bubbles popping. The hissing of a wave as it retreats seaward is perhaps caused in part by sand grains rubbing against each other.

An important part of scientific thinking is to doubt the theories proposed by others. Do you think the explanations given above are reasonable or not? Are they similar to the hypothesis you made in the previous activity? Can you devise a test to prove or disprove the explanations?

Longshore Currents

Nothing in the world is more flexible and yielding than water.
Yet when it attacks the firm and the strong, none can withstand it.

—Lao Tzu (Chinese philosopher and poet)

It's a hot summer day at the beach. You decide to cool off in the water. For a long while you float in the surf, delighting in the warm glow of the sun, the cry of gulls, the sound of waves pounding sand. When you finally look back to the beach, you're shocked at how far you've drifted from where you first entered the water. What happened?

What you experienced is the longshore current (also called longshore drift). If you watch the waves coming in, you can see how longshore currents form. When waves strike the shore at an angle, they sweep along the beach, pushing water laterally, creating a current that flows parallel to the shore.

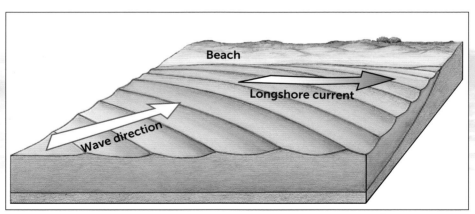

Longshore currents form when waves strike a beach at an angle, pushing water down the length of the beach.

The direction of a longshore current varies from time to time depending on the direction of the winds that form the waves. The usual flow of longshore currents in the Carolinas is from north to south (or west to east in the case of Shackleford Banks). This is especially easy to see in the winter when strong winds often blow in from the north. In the summer, winds usually come from the south and longshore currents generally flow to the north.

Longshore currents can be dangerous. During a storm, they can knock people off their feet and push them onto pier pilings. But longshore currents move more than swimmers. They also move sand up and down the beach and by doing so, help determine the shape of the beach.

How Fast Is the Longshore Current?

A technique favored by oceanographers worldwide for measuring the speed of a longshore current is to toss an orange into the surf and see how fast it moves. Why an orange? Because the orange stays partly submerged and so is unaffected by wind. It's also brightly colored and therefore easy to see. And best of all, when the calculations are done, the hungry scientist can retrieve, peel and eat the orange.

WHAT YOU NEED An orange, a calculator (or just a brain), a tape measure and a timepiece.

WHAT TO DO

1. Note which way bubbles, froth or driftwood are moving in the surf. Or look at the angle the waves are making to the beach. That tells you the direction of the longshore current.

2. Longshore currents are often measured in feet or meters per second. To measure current speed in miles per hour, draw a line on the beach perpendicular to the water. This will be your starting line. Draw another line 53 feet (roughly 1/100 of a mile) down the beach in the downstream direction of the longshore current. That will be your finish line.

3. Toss an orange into the surf zone. Record how many seconds it takes the orange to drift from starting line to finish line.

4. Calculate the current velocity. If T is the time in seconds it takes an orange to drift 53 feet, then 36 / T = velocity of a longshore current in miles per hour. For example, if it took 30 seconds for the orange to reach the finish line, then the velocity of the current can be calculated as follows: 36 ÷ 30 = 1.2 mph. Can you figure out how the equation was derived? See the glossary for help.

Aside from their effect on swimmers, why are longshore currents important? How do you think wave size affects longshore current speed? How does the angle that a wave approaches a beach influence the current's strength? How much sand can a longshore current transport in one year and where does all that sand eventually go?

More about Longshore Currents

Longshore currents are the main force that moves sand along the beach. For instance, at Nags Head, North Carolina, about 500,000 cubic yards of sand moves south every year. That's about 50,000 dump truck loads of sand! Actually, more sand than that moves south every year, but in the summer some sand also moves to the north. The overall net flow of sand is calculated by subtracting the summer volume from the winter volume. Under the right conditions, such as during a storm, longshore currents can move so much sand that they are a major cause of beach erosion. Some of the sand gets dumped in inlets, some relocates to another beach on a nearby island and some piles up against groins and jetties (see Activity 32).

The speed (or strength) of a longshore current increases if

1 the waves get bigger
2 the beach slope gets steeper
3 the angle of wave approach gets larger

EXPLORE YOUR WORLD

Measure the speed of the longshore current under different tides, wave conditions and wind conditions and if possible, in different seasons. If you are feeling really ambitious, go to an inlet and do this same activity with an orange to measure the velocity of the tidal current (see Activity 6). *Be careful!* Tidal currents can be swift. If something drifts out to sea, make sure it's the orange, not you!

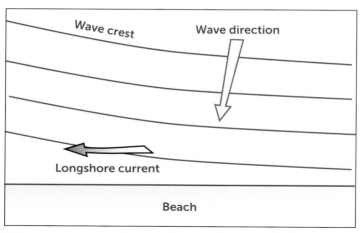

When waves approach a beach at a slight angle, the longshore current is weak. That's because most of the wave energy bounces off the beach.

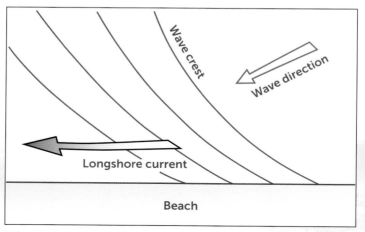

When waves approach a beach at a large angle, the longshore current is strong. That's because much of the wave energy is directed laterally along the beach.

Rip Currents

Only a fool tests the depth of the water with both feet.

—African proverb

Rip currents are narrow streams of water that flow perpendicularly to the shoreline. In the Carolinas they can happen on any ocean beach, especially when the waves are high. They seem most abundant on the Outer Banks, where high waves are common.

Rip currents (sometimes mistakenly called riptides) form when water forced on shore by breaking waves escapes in a single offshore moving stream. The location of rip currents may be controlled by groins, jetties, sandbar gaps, piers or any place where two longshore currents, moving in opposite directions, meet head on. When longshore currents bump up against groins or jetties, the water is forced out to sea, sometimes at a rapid velocity. Rip currents are very dangerous to swimmers, who can be dragged far from the shore. They can occur suddenly and without warning, although expert lifeguards will often warn swimmers if rip currents are likely to occur.

Rip currents can pull the unwary swimmer a hundred yards or more out to sea.

How to Identify Rip Currents

Rip currents are usually visible to the wary beach stroller and appear as a stream of seaward moving water flowing through the line of breaking waves. Other signs of a rip current include:

- A reduction in **wave height** where rip currents flow through the surf
- A change in the color of the water
- A rapid seaward movement of foam, seaweed or driftwood

If you suspect the presence of a rip current but are unsure, toss some seaweed or driftwood into the surf and see what happens. That will tell you something about the speed and direction of the current.

WHAT YOU NEED A good pair of eyes and a pier (optional) or a seaside hotel room (also optional).

WHAT TO DO This is an observational activity particularly for a day of relatively high waves. Walk up and down a beach. Watch for rip currents and note their appearance and location. Rip currents are easier to spot from a height. All the better, then, if you can get out on a pier and stand by the railing over the zone of breaking waves. Come back to the same location and observe the rip current again, at a different tide stage or on a different day. Is the rip at the same location? Does it have the same appearance? Can you see a reason (for example a gap in an offshore sandbar) for the location of the current?

WHO LIKES RIP CURRENTS?

Kayakers and surfers often seek out rip currents because they can provide an effortless ride out into the ocean.

How to Escape a Rip Current

Rip currents are responsible for perhaps more than 100 drownings every year in the United States and 80 percent of all lifeguard rescues at American beaches. When caught in a rip current, *do not panic*. Wave your arms and shout for help to let those on shore know you are in trouble. Then either 1) swim parallel to the shoreline until you are outside the rip or 2) float with the current out to sea to where the rip ends. Then swim toward the side of the current and back to land. *Do not swim toward land* until you have escaped the narrow stream of water in a rip. Rip currents flow from one or two feet per second to as fast as eight feet per second. That's faster than an Olympic swimmer!

Some rip currents last only a few hours, disappearing after wave, wind or tide conditions change. Others have been known to last for weeks. Strong, persistent rips can be a major cause of local beach erosion. Myrtle Beach, South Carolina, is the deadliest beach in the Carolinas for rip current fatalities.

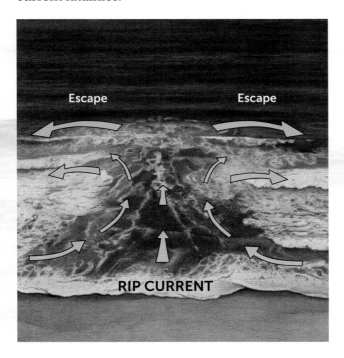

Tides

The tide rises, and the tide falls,
The twilight darkens, the curlew calls
—Henry Longfellow
(nineteenth-century American poet)

Tides are the daily up-and-down changes in **sea level** caused by the gravitational pull of the moon and the sun as the earth spins. Since the moon is closer to the earth than the sun, it has a much greater effect on the tides. The moon's gravitational pull causes ocean water to pile up or bulge on the side nearest the moon. At the same time, another tidal bulge or high tide forms on the opposite side of the earth. High tides occur as the earth rotates on its axis and a shoreline reaches one of the tidal bulges. In most places worldwide, including in the Carolinas, this happens twice a day, so there are two high tides and two low tides each day.

Tidal range is the vertical distance between high and low tide. Daily tidal ranges for locations along the coast are listed in **tide tables** (or tide charts) found online or in local newspapers.

Who uses tide tables? Sea captains navigating shallow harbors, fishermen seeking certain kinds of fish, meteorologists trying to predict storm damage and surfers hoping to catch the ultimate wave all make use of tide tables.

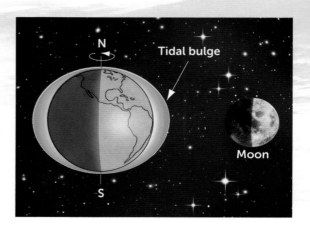

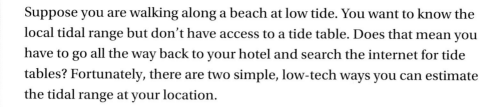

Suppose you are walking along a beach at low tide. You want to know the local tidal range but don't have access to a tide table. Does that mean you have to go all the way back to your hotel and search the internet for tide tables? Fortunately, there are two simple, low-tech ways you can estimate the tidal range at your location.

The Barnacle Method

Barnacles attach themselves to just about any solid surface, including rocks, pilings, ship bottoms and even other marine life. They are **filter feeders** (animals that filter out nutrients from seawater) and so cannot live for long out of the water and away from their food source. Usually found in the **intertidal zone**, barnacles can be used to get a rough estimate of local tidal range.

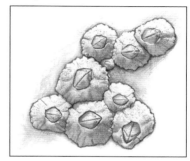

Acorn barnacles are a common sight on Carolina piers.

WHAT YOU NEED A pier and a tape measure.

WHAT TO DO Visit any pier at low tide when its barnacle-encrusted pilings are exposed. Find a piling close to the edge of the water and measure the vertical distance from the beach to where the barnacles disappear and are replaced by marine algae. That distance will tell you roughly what the tidal range is over the course of a month at that particular beach. Unfortunately this method only works on beaches with piers and tells you nothing about what the tide is doing at a particular time or on a particular day.

The Stick Method

WHAT YOU NEED A stick (yardstick, driftwood, etc.), a tape measure and a partner.

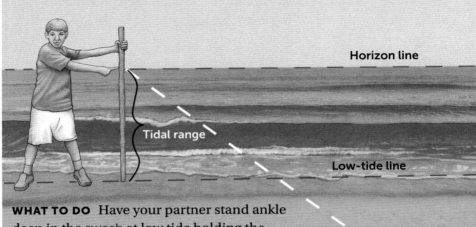

Horizon line

Tidal range

Low-tide line

WHAT TO DO Have your partner stand ankle deep in the swash at low tide holding the stick upright. Lay your head flat on the sand on the high-tide line (usually a couple of feet seaward of the wet/dry line). Sight on the imaginary point where the horizon line meets the stick. Direct your partner to mark that point on the stick. The distance from sand to mark should be close to the tidal range for that particular location and time. Compare your estimate to tide table predictions.

High-tide line

Wet/dry line

Problems Estimating Tidal Range

Sometimes tidal range estimates are different from tide table predictions. Why do you think this is so? Listed below are several factors that can affect estimates of tidal range.

WAVES

Locating the high-tide line may be difficult if the surf is big. Why? Because big waves may push the **wet/dry line** farther up the beach away from the normal high-tide line, making your estimates larger that the actual range.

TIDE GAUGE LOCATION

Tides are measured at **tide gauges**, often miles away from where you are on the beach. Tidal ranges at your beach may differ from predictions based on tide gauge measurements.

WIND

Strong winds blowing **onshore** (from sea to land) can cause tides to be higher than predicted. Likewise, a strong **offshore wind** causes tides to be lower. Under extreme conditions, such as during a hurricane or a **nor'easter**, tides in the Carolinas can exceed 10 feet above the normal range.

Did you know?
The world's largest tidal range is 55 feet at the Bay of Fundy in Nova Scotia, Canada.

RAIN

Heavy local rains or rainwater delivered by rivers to coastal regions can increase tides by a foot or more, especially at beaches near rivers.

ATMOSPHERIC PRESSURE

The column of air over the sea has weight that pushes down on the sea. But that weight varies with changing weather. A high-pressure system brings denser, heavier air that might make the tide lower than expected. A low-pressure system brings less dense, lighter air that can create higher than expected tides.

Spring Tides, Neap Tides and King Tides

When the combined gravitational pull of the moon and the sun is at a maximum, tides are also at a maximum. This is referred to as a spring tide and occurs during a full moon and a new moon. Spring tides are higher than normal high tides and lower than normal low tides. When the moon is in the first or third quarter, the tidal range is at a minimum and is called a neap tide.

   A king tide is a recently coined term for exceptionally high spring tides. King tides occur a few times each year when both moon and sun are closest to the earth. Both spring tides and king tides are important because they can cause widespread coastal flooding, especially if they coincide with a major storm. The 1962 Ash Wednesday Storm, a nor'easter that killed 40 people and damaged businesses, homes and motels from New Jersey to the Carolinas, occurred during a spring tide. It is considered to be one of the ten worst storms to hit the U.S. East Coast in the twentieth century because of strong winds and flooding that lasted for five consecutive high tides.

SPRING TIDES

Full
Moon

Earth

New Moon

Sun

High Tides

1st Quarter Moon

NEAP TIDES

High Tides

Earth

Sun

3rd Quarter Moon

More about Tides

Tides are really giant waves that circle the world as it spins. When the crest of the wave arrives, it's high tide. The trough of the wave is low tide. This wave that we call the tide creates tidal currents, though we don't usually feel them at the beach. But stand at a harbor dock or at the edge of an inlet and you will see a tidal current, sometimes moving at an impressive clip.

The intertidal zone, with its alternating periods of wet and dry, is a harsh environment for life. But beach critters have evolved techniques to handle such conditions. Some, like clams and mole crabs, migrate up and down the beach with the tides. Other species, including some fish, mollusks and sea turtles, often lay their eggs on a beach when the high tide reaches its greatest extent so their eggs won't be drowned by the next high tide.

In addition to the wind, the tidal range is affected by the width of the continental shelf. A wider continental shelf means a greater tidal range. That's because the continental shelf acts as a wedge that pushes up the incoming tide. The wider the continental shelf, the more the tide wave is pushed up. At Duck, North Carolina, where the continental shelf is 30 miles wide, the average tidal range is 3.2 feet. By contrast, on Hilton Head the range is 7.4 feet. There the continental shelf is 80 miles wide (see Glossary).

North Carolina	Normal tide (ft.)	Spring tide (ft.)	South Carolina	Normal tide (ft.)	Spring tide (ft.)
Duck Pier	3.2	4.0	Myrtle Beach	5.0	6.0
Cape Lookout	3.7	4.4	Murrells Inlet	4.3	4.7
Cape Hatteras	3.4	4.1	Pawleys Island	5.0	5.8
Atlantic Beach	3.6	4.3	Folly Island	5.2	6.1
Yaupon Beach	4.8	5.5	Hilton Head	7.4	8.5

2

Moving Beaches

You cannot step in the same river twice.
—Heraclitus

Heraclitus was an ancient Greek philosopher. Only fragments of his writings survive, but it seems clear he believed everything in the world to be in a state of constant change. Everything changes, he thought, because nature consists of opposing forces that are always in conflict with each other. Conflict between opposites, Heraclitus claimed, paradoxically creates the unity we see in nature. Today we might call that unity the ecosystem.

A beach is an excellent example of conflict between opposing forces. Wind pushes against water, forming waves. Waves move sand around. Plants rooted in the sand resist both waves and wind. People build structures like seawalls to oppose waves. Sea level rises and enhances the power of those waves. These forces in opposition are why beaches are constantly moving, changing, evolving, becoming something different. Moreover, the same forces moving beaches around today have been in operation for hundreds of millions of years and will continue operating far into the future. If Heraclitus were alive today, he might well say, "You cannot step on the same beach twice."

Sea Breeze

You don't need a weatherman to know which way the wind blows.
—Bob Dylan (American singer and songwriter)

Mother and daughter stood together on an empty beach, watching the waves come thundering in.

"How are beaches formed?" asked the daughter.

"Waves make beaches," answered the mother, "by pushing sand around."

"So what makes the waves?"

"The wind, of course."

"Why does the wind blow?"

"Heat from the sun," said the mother, "causes the air to move around."

"How come the sun is hot?"

"Hydrogen atoms fuse together, giving off light and heat. It's kind of like a nuclear bomb," explained the mother.

"Why do the atoms fuse together?"

"Gravity."

"Oh, so the sun's gravity makes beaches. What makes gravity?"

"You ask a lot of questions," said the mother. "Let's go for a swim."

Everything in nature is connected by an intricate chain of cause and effect, though we don't always see all the links in the chain. It's easy to stand on a beach, for example, and watch waves moving sand around, but we tend to forget it's the wind that makes the waves. In that sense, wind is the fundamental cause of major changes on a beach.

Sea Breeze

When the breeze blows from sea to land, it's called an **onshore wind** or, simply, a sea breeze. When the wind blows from land to sea, it's known as an **offshore wind** or land breeze. During conditions of fair weather, sea and land breezes are nearly always present at the beach. What's interesting is how these breezes predictably change every day.

WHAT YOU NEED Access to a beach during the day and at night.

WHAT TO DO Visit the beach during the day (midafternoon is best) at a time of fair weather when no storms are passing through. Note the direction and strength of the wind. Is it an onshore or offshore breeze?

Return to the same beach late in the evening or very early in the morning before sunrise. Where is the wind coming from? Is the direction of the wind at night the same during the day? What about wind strength?

Repeat this activity over the course of several days. How does the wind change in direction and strength everyday? Why? What effect does the land/sea breeze have on waves, currents or the plants beyond the dunes?

During the day, warm air rising over land is replaced by cooler air coming from the sea. This creates a cool breeze known as a sea breeze or an onshore wind.

warm air rising

Sea breeze

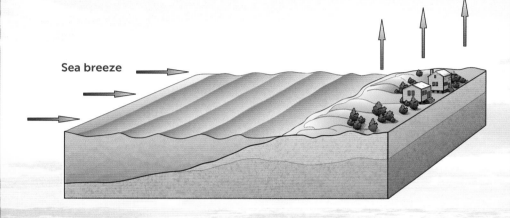

At night, warm air over the sea rises and is replaced by cooler air coming from land. This creates a land breeze or an offshore wind.

warm air rising

Land breeze

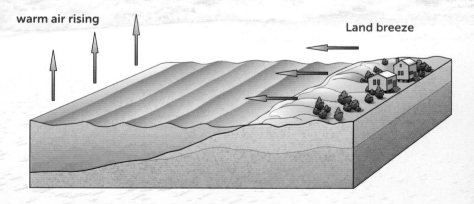

More about the Sea Breeze

During the day the wind comes in from the ocean, while at night it flows from land to sea. This daily reversal in wind direction happens because land gets heated by the sun faster than water does. As the land gets warmer, it heats the air above it. The air expands, becomes less dense and rises. Cooler, denser air above the nearby sea sinks and moves landward.

At night it's the opposite. The sea is warmer because it retains heat from the sun longer than does land. The air above the sea, being warmer and less dense, rises. Cooler, denser air from land moves in to replace the rising sea air. Because the temperature difference between land and sea is greatest during the day, the wind is generally stronger during the day.

The sea breeze affects beaches in several ways:

- It increases **wave height**, which in turn strengthens **longshore currents** (Activity 4).
- It affects the **tidal range** (Activity 6) by piling up seawater along the shore. Land breezes push water away from shore.
- The sea breeze shapes plants by blowing salt spray into the dunes and **maritime forests** (see Activity 29).

Wind direction has seasonal as well as daily changes. Summer winds on the east coast generally come from the southwest. In the winter, winds often come from the northeast. Wind direction determines the direction of the longshore current.

DID YOU KNOW? ───────────────────────

Sea breezes in the summer bring cool, moist air from the water into the **coastal plain**. As the air rises over warm land, it cools and condenses, forming clouds. That's why you see a lot of clouds over land in the afternoon and why afternoon thunderstorms are so common.

The Beaufort Wind Scale (at the Beach)

In 1805 British admiral Francis Beaufort devised a wind scale to help sailors estimate wind speed. The scale, based on the observed effects of wind on the sea and on land, is still used by mariners today, though in a modified form. Use the Beaufort Wind Scale to estimate wind speed at your beach. Compare your estimates with the wind conditions reported by local news channels. What do you see happening on the beach as the wind speed increases?

Beaufort number	Description	Wind speed (mph)	Conditions
Force 0	Calm	0–1	Smoke rises vertically; flat, glassy sea
Force 1	Light air	1–3	Ripples without crests, flags limp
Force 2	Light breeze	4–7	Small wavelets (8 in.), trees rustle slightly, wind felt on face
Force 3	Gentle breeze	8–12	Large wavelets (2 ft.), a few whitecaps, light flags extended
Force 4	Moderate breeze	13–18	Small waves (3 ft.), numerous whitecaps, small branches moving, light flags extended
Force 5	Fresh breeze	19–24	Waves 4–8 ft., many whitecaps, some spray, flags snap, beach sand moves
Force 6	Strong breeze	25–31	Waves 8–13 ft., more spray, whitecaps everywhere
Force 7	Near gale	32–38	Big waves, trees moving, foam blown in streaks, sand stings face
Force 8	Gale	39–46	Waves start to break up in spray, walking difficult, waves 18 ft.
Force 9	Strong gale	47–54	Dense foam, spray in air, tree branches breaking
Force 10	Storm	55–64	Huge waves, spray everwhere, sea white, little visibility, foam in air
Force 11	Violent storm	64–73	Foam covers sea, visibility further reduced
Force 12	Hurricane	73+	High waves with overhanging crests, visibility reduced, uprooted trees

Dunes

Sea waves are green and wet,
But up from where they die,
Rise others vaster yet,
And those are brown and dry.

—Robert Frost (a popular twentieth-century American poet)

Dunes are those magnificent mounds of sand that rise at the edge of the beach like waves of grass-crested earth. An inspiration for generations of painters and the stuff of bloody legend from the history of warfare,* dunes are also a wildlife habitat, a protective barrier against storms and a sand supply to sustain beaches during times of coastal erosion.

Dunes are children of the wind. They are created, shaped and moved by wind. As it marches sand grains across the beach, the wind builds up a dune, grain by grain. Dunes constantly change shape and move with shifting winds. Plants like **sea oats** can keep a dune in place, slowing down its advance. And sometimes trees establish themselves over a dune field and the dunes stay put for years, even centuries, until some new force comes into play, a major storm perhaps or human intervention. Then the wind once more gathers sand and the march of the dunes begins anew.

Coastal landforms are not always what they seem. Many dunes on Carolina beaches that appear to be natural are actually artificial piles of bulldozed sand. Even though a dune is formed from beach sand, its sand grains are visibly different from the sand grains at the beach. And the layers of sand that make up a dune are different from those that make up a beach.

*Dunes have been a point of attack for invading armies since ancient times. Famous examples include the Allied landings at Normandy in World War II and the many battles fought over American Civil War forts built behind or in the dunes.

How can you tell if a dune is natural or made by a bulldozer? How is dune sand different from beach sand? Why are dunes ecologically important, and how does a simple process like wind blowing sand create forms of such complexity and beauty?

The Birth of a Dune

Dunes are made from wind-blown sand. But something blocks the wind and allows sand to pile up. Your task is to find what that something is.

WHAT YOU NEED A beach with dunes and a notebook.

WHAT TO DO Stroll along the beach and examine the sand. Record in your notebook such things as grain size, numbers of shell fragments, sand color and sorting. Are all the grains the same size (well sorted) or are they different sizes (poorly sorted)? Now walk above the high-tide line to the edge of the dunes. *Take care to minimize your effect on the dunes and especially avoid trampling the vegetation.* Examine the sand on the dune

Wind

Dunes have a steep side that faces away from the dominant wind direction and a gentle (less steep) side that faces into the wind.

surface. How does it compare to beach sand? Record your observations in your notebook.

Walk along the upper beach right in front of the dunes. You should see some small dunes starting to form. These "baby" dunes can be a few inches to a few feet in height. Ask yourself why a baby dune is forming at a particular spot. Can you see anything around which the sand is gathering?

How to Make a Dune

Dune sand has smaller grains and is better sorted than beach sand. The larger, heavier grains (including shell fragments) drop out of the wind as it moves sand up the beach, so only the finer grains make it up to the dunes. That means if you find shells and large, poorly sorted grains in a dune, you're looking at a bulldozed pile of beach sand, *not* a natural dune.

Dunes are created when **wrack** (driftwood, seaweed, plastic and other debris washed ashore) forms a nucleus around which sand accumulates. If you looked closely, you should have seen small piles of sand sheltered in the downwind side of these objects. Given time, these piles may become genuine dunes.

Examine an eroded dune face or make a small cut with a shovel (to be filled in when you are finished) and you will see very thin layers of sand (beach sand has thicker layers). Some layers may be tiny brown shell fragments. Others may be black **heavy minerals** or clear **quartz** grains.

Many species of plants and animals live in **symbiosis** with each other (they help each other) and with the dunes themselves. Sea oats, palmettos and a score of other dune plants provide food, shelter and nesting sites for birds, mammals and insects. Plant roots stabilize the dunes and dunes shelter the organisms living there from storms. Dunes also shelter houses (though not from a major storm).

The birth of a dune, Edisto State Park, S.C. Dunes, like the ones in this photo, are formed when sand collects around an object, usually something from the wrack. If enough sand is gathered, plants like sea oats may colonize and stabilize the sand pile. This allows more sand to gather, and the dune grows ever larger.

At over 100 feet high, Jockey's Ridge at Nags Head is the tallest dune on the East Coast.

Ancient Shorelines

Geology gave us the immensity of time and taught us
how little of it our own species has occupied.
—Stephen J. Gould (American paleontologist and science writer)

The beaches of the Carolinas were not always where they are now. Over the course of millions of years our shoreline has marched back and forth across the coastal plain in response to changing sea levels. When sea level was high, the shoreline moved west, sometimes flooding the land all the way to the Piedmont. When sea level was low, the shoreline shifted east, sometimes moving out over the continental shelf. Dry land became the bottom of the sea and land that had formerly been under water became dry land again.

Near the end of the Pleistocene (18,000 years ago—see the geologic time scale at the end of the book), sea level was 400 feet lower than at present. The beaches of the Carolinas lay about 40 or 50 miles east of where they are today. The ocean where you swim today was then an Ice Age forest where mastodons and saber-toothed cats prowled.

The mastodons are gone, but you can still find the remains of Pleistocene shorelines scattered across the coastal plain. These "fossil" shorelines are a series of long, thin ridges that geologists in the 1920s identified as old shorelines. They called them terraces or scarps but didn't know how they formed. Now geologists recognize that these ridges are the remains of old barrier islands (much like the islands off the Carolinas today; see Activity 11) left behind when sea level dropped.

Searching for Lost Shorelines

WHAT YOU NEED A car, a licensed driver and a good pair of eyes.

WHAT TO DO As you drive to or from the beach, look for evidence of ancient shorelines. Scan the landscape for small sandy hills or ridges that appear rather suddenly. If they are oriented in a southwest to northeast (sometimes north-south) direction and are composed of light colored sand, then they are almost certainly old barrier islands. Basically, any prominent sandy ridge on the coastal plain is the eroded leftover of a barrier island. These islands once hugged the old shorelines of the East Coast, much like barrier islands do today.

During the warm periods of the Pleistocene, the world's glaciers and ice sheets melted and added their waters to the sea. Sea level rose and the barrier islands migrated landward, just like they are doing today. During the cold periods, a lot of the world's seawater got tied up in the glaciers. This caused sea level to fall, leaving the barrier islands high and dry, orphaned by a long departed sea.

Barrier islands are created when sea level rises (see Activity 11). When sea level falls, the islands stop migrating. Land emerges from the receding ocean and the islands stay behind as unmoving hills of sand on an otherwise level coastal plain.

The Orangeburg Scarp near Laurel Hill, N.C., is the low ridge cutting across Highway 74 in the distance.

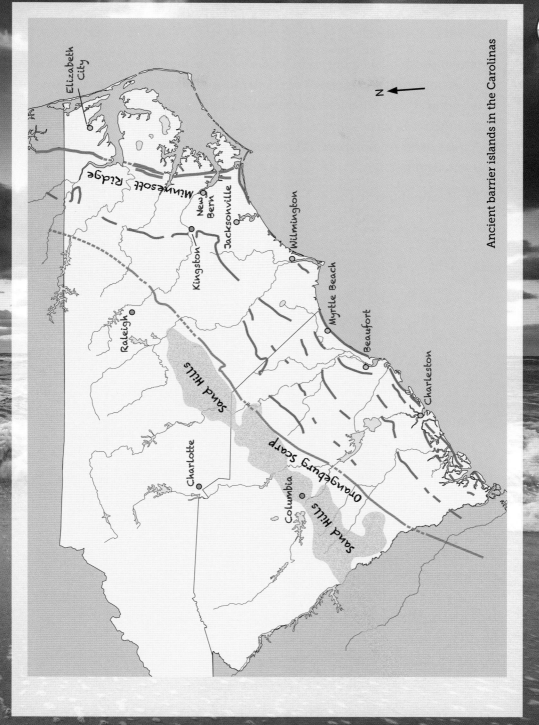

Ancient barrier islands in the Carolinas

Elizabeth City

Minnesott Ridge

New Bern

Jacksonville

Kingston

Wilmington

Raleigh

Myrtle Beach

Sand Hills

Beaufort

Charleston

Charlotte

Orangeburg Scarp

Columbia

Sand Hills

N

Ancient Islands

As you discover the ancient shorelines of the Carolinas, try to imagine what the land and sea must have looked like thousands and even millions of years ago. What will they look like a thousand years in the future? Listed below are places where you can see old barrier islands. There are hundreds more from Florida to Virginia, some yet to be discovered. Happy hunting!

The Orangeburg Scarp. A chain of two-million-year-old barrier islands stretching from Florida to northern Virginia. You can see the scarp where it crosses Highway 74 near Laurel Hills, North Carolina, and near the intersection of Highway 601 and the Congaree River in South Carolina.

Havelock, North Carolina. Parts of a 125,000-year-old barrier island (the Minnesott Ridge) cross Highway 70 in Havelock. Go north on Highway 306 and take the ferry to Minnesott Beach to see a well-preserved cross section of the old island. As sea level was rising, the island migrated west and buried an old cypress swamp, so now you can see 125,000-year-old cypress stumps at the edge of the Neuse River. You can also see the Minnesott Ridge exposed at Flanner Beach, a few miles north of Havelock near the town of Croatan.

Cities on top of old barrier islands. Myrtle Beach in South Carolina and Wilmington, Swansboro, Morehead City and Beaufort in North Carolina were all built on top of old barrier islands.

EXPLORE YOUR WORLD

Go online to find out more about the plants and animals of the Pleistocene. Do a drawing in your sketchbook of what you imagine the Pleistocene landscape must have looked like in eastern North and South Carolina.

Beach Erosion

> I have seen the hungry ocean gain
> Advantage on the kingdom of the shore,
> And the firm soil win of the wat'ry main,
> Increasing store with loss, and loss with store.
> —Shakespeare, Sonnet 64

Beach erosion is the removal of beach sediments (mostly sand in the Carolinas) by wind and water. Nearly all of the beaches in North and South Carolina are eroding. Why is that? What causes beach erosion? How do you know for sure that a particular beach is eroding? What about the rate of erosion? Can you tell if a beach is eroding rapidly (10 feet or more per year) or slowly (one or two feet per year)?

How to Know If a Beach Is Eroding

WHAT YOU NEED A strong sense of curiosity.

WHAT TO DO Walk down the beach for a mile or so (best to do this at low tide) and try to answer the following questions:

Is this a **replenished** beach? (See Activity 31.)
Are **seawalls** or **groins** present?
Do you see **scarps** (small cliffs) on the beach?
Are there dune walkovers without a dune?
Is the high-tide mark right at the base of the dunes?
Is the beach/dune boundary a sand cliff?
Are there houses seaward of the high-tide line?
Are there trees or stumps on the beach or in the water?
Do local people talk about an erosion problem?

If the answer is yes to all or to most of the questions, then the beach is eroding. Groins, seawalls and **beach replenishment** are put in place to combat erosion, so they are sure signs that a beach is eroding. Likewise, if something is where it shouldn't be (houses, pilings and dune walkovers), erosion may be the cause.

Obviously, trees don't belong on a beach and must be there because the beach has moved. Tree stumps still rooted to the ground (the remains of drowned forests) are a common sight on the Outer Banks, usually showing up after a storm. In South Carolina you can see spectacular examples of dead trees standing on the beaches at Hunting Island and on Bull Island.

Sand cliffs at the beach/dune boundary, beach scarps and dead trees are signs of rapid erosion, but they don't tell you how fast a beach is eroding. To determine the erosion rate, scientists study old maps, satellite imagery and photographs and talk to longtime residents or visit a particular beach over the course of several years. Scientists also note how beaches, like the example of Morris Island (page 54), shift through time relative to any buildings constructed on them.

Beach scarps like this one at Edisto Beach State Park, S.C., are signs of beach erosion.

What Causes Beach Erosion?

There are many causes of beach erosion. Storm waves can move an entire beach in a single day. Even small waves during a spring tide can cause major erosion. Wind, though not as powerful an erosive force as water, can still remove a lot of sand. And under the right conditions, longshore currents can take away massive volumes of sand. Of course a natural beach never completely disappears. It just relocates.

Humans also play a major role in coastal erosion. Engineering projects like seawalls or groins may forestall local erosion (at least temporarily), and a jetty may keep a shipping channel open. But these projects often create problems at other locations. A good example is the Morris Island Lighthouse near Charleston, South Carolina. Built in 1876 on the sound side of the island, the lighthouse was surrounded by 15 buildings, including quarters for the light keeper and a one-room schoolhouse.

The Morris Island Lighthouse, near Charleston, S.C.

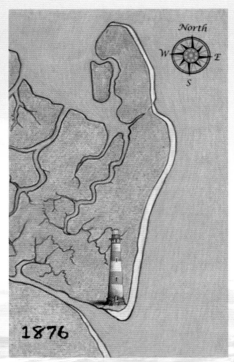

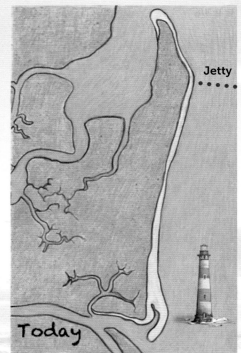

Morris Island in 1876 and today

In 1889 jetties were constructed to maintain the entrance to Charleston Harbor. The jetties kept the harbor channel open but blocked the sand supply to Morris Island coming from the north, causing the island to erode. Every year the sea came a little closer to the lighthouse. By 1938 it was at the edge of the surf. Today the lighthouse stands alone, surrounded by water, several hundred yards from land.

Shoreline Retreat

The sea is rising on the shores of the world, marching relentlessly inland like an unstoppable army, without mercy, without pause. Beaches are moving back, or "retreating," as they have done countless times in the past. Because it more accurately describes how beaches respond to sea level rise, scientists prefer the term "shoreline retreat" in place of coastal erosion.

Dead trees on a beach, like these on Hunting Island, S.C., are the victims of erosion. Salt from the encroaching sea kills the trees, creating the so-called boneyard beaches.

Barrier Islands

And thus through many seasons' space
This little Island may survive.

—Dorothy Wordsworth (poet and the sister of poet William Wordsworth)

Barrier islands are long, narrow strips of sand with inlets at either end. Nearly all ocean-facing beaches in the Carolinas are sitting on barrier islands. Each island is unique but usually consists of four major zones: 1) beach, 2) dunes, 3) maritime forest and 4) salt marsh.

Barrier islands are special. They have the unique ability to move toward the mainland as sea level rises. This process is called barrier island migration. As we learned in Activity 9, barrier islands have been migrating for millions of years. How is this possible? How can an entire island move? How do we know for sure that barrier islands migrate?

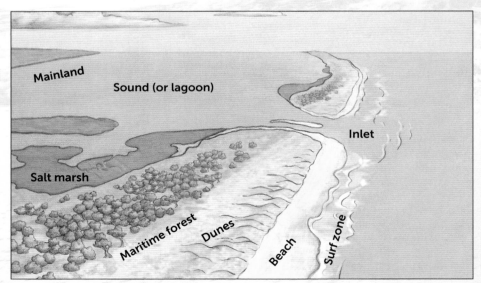

A typical barrier island in the Carolinas.

Prove Barrier Islands Migrate

As a barrier island moves toward the mainland, the beach (given enough time) will eventually be right on top of where the salt marsh used to be. When that happens, you may see parts of the marsh exposed on the beach, providing evidence that the island has moved.

WHAT YOU NEED Access to a barrier island beach, preferably one that has not been replenished (not had sand artificially pumped in).

WHAT TO DO Walk along the beach and look for the following: oyster shells, marsh periwinkles, peat (partly decayed salt marsh plants), tagelus clams (see Activity 21) and mud. All of these come from the salt marsh. Finding them on a natural beach tells you the island has migrated.

Marsh periwinkles (left) and oysters (right) live in salt marshes.

How Islands Migrate

The reason barrier islands migrate is simple—rising sea level. The mechanism for their migration is storms. It works like this:

1 A powerful storm like a hurricane or nor'easter hits a barrier island. Waves wash sand through gaps in the dunes. Sand that used to be on the beach is carried across the island through gaps in the dunes (a process known as overwash).
2 Year after year more storms come. More beach sand is washed over the island. Sometimes the sand is deposited on the middle of the island. Sometimes it is dumped in the sound on the back side of the island.

3 Over time the beach erodes and moves back. The middle of the island gets higher. The back side of the island widens and moves back. The result is that the barrier island moves up (gets higher) and moves back. Ironically, the beach erosion that worries so many people is what allows barrier islands to survive sea level rise.

As the sea rises, barrier islands migrate toward the mainland. At the same time, the shoreline of the mainland is also retreating. So the islands, more or less, keep the same distance from the mainland. But sometimes sea level rises too fast and islands are overrun by sea level rise. And sometimes the islands migrate too quickly and merge with the mainland or get left behind when sea level drops.

Overwash fan

Satellite image of Core Banks, N.C., showing an overwash fan where a storm carried sand from the ocean side to the sound side of the island. Year by year, storm by storm, this is how barrier islands move toward the mainland. (See Activity 39 to learn how to use Google Earth to see overwash fans and other cool things.)

A Thought Experiment

A thought experiment uses the imagination to find out things about the world when a physical experiment is impractical. The following thought experiment is about shoreline retreat.

Sullivan's Island is a developed barrier island near Charleston with expensive houses, stores and motels. A few miles to the north is undeveloped Bulls Island, home to deer, alligators, birds and other wildlife. Scientists expect a 3-foot sea level rise by the end of the century. How will the inhabitants on these two islands likely respond to sea level rise? What will the islands look like in 50 years?

Initially, both islands will try to migrate to the west. But the migration will be prevented on Sullivan's Island. People will try to hold the island in place to save expensive homes. They will pump sand on the beach and build ever-higher seawalls. But because seawalls eventually destroy beaches (see Activity 32), the beach on Sullivan's Island will be entirely lost (probably within 30 years). Bulls Island, on the other hand, will move westward. The beach will remain wide and healthy (though with many tree stumps). The alligators and deer won't mind. They'll just move back with the island.

This thought experiment raises an important question: Is it better to destroy a beach in order to save houses, or is it better to let the sea destroy houses and save the beach by moving back like the deer and alligators? What do you think?

Peat deposits at Edingsville Beach, S.C. Peat is a mix of mud and partly decayed plant and animal remains. It forms in salt marshes on the back (sound side) of barrier islands and also in freshwater marshes (swales) in the middle of barrier islands. When peat shows up on the ocean side of a barrier island, that means the island has moved and is now sitting on top of an old marsh. Note the oysters pointing up, in the same position as when they were alive. You probably don't want to swim at this beach!

Most of the peat you see on Carolina beaches comes from spartina plants growing in salt marshes like the ones near Hilton Head, S.C. (pictured). Notice the marsh periwinkles crawling on the spartina.

CHAPTER

3

Sand

In every grain of sand there is the story of the Earth.
—Rachel Carson
(American marine biologist and conservationist)

Not all beaches are sandy. Some are made of boulders, pebbles, shells, mud or coral. Beaches in the southeastern United States are made of sand, with a smaller percentage of shell fragments and organic matter (driftwood, seaweed, etc.). The percentage of shell material in beach sand increases the farther south you go, reaching a peak in southern Florida.

For most beachgoers, sand is a handy material for building castles or perhaps for a warm surface upon which to sunbathe. But sand has an allure beyond its recreational value. Look closely at grains of beach sand and you will see tiny crystals of tourmaline, garnet, topaz and other semiprecious gemstones. Each grain has a fascinating story to tell about a journey spanning hundreds of miles and lasting millions of years. Grab a handful of the beach sand and you are grabbing a piece of the mountains, for that's where sand grains originally came from.

The activities in this chapter describe the composition, origin and size of sand grains and how to identify the various minerals found on a Carolina beach. The chapter also resolves two great mysteries that have puzzled scientists for centuries: why sand barks like a dog when you walk on it and why sand is sometimes so soft you sink up to your ankles.

Sand

To see a World in a Grain of Sand

And a Heaven in a Wild Flower,

Hold Infinity in the palm of your hand,

And Eternity in an hour.

—William Blake (English poet and artist)

Beach sand has two main components: minerals and shell fragments. The shell component is called the carbonate fraction. It consists of shells broken by waves, cars and by predators who crack open shells to harvest the inhabitants (see Activity 30). In the Carolinas, the carbonate fraction varies widely from beach to beach but usually makes up about 5–25 percent of the sand.

The main mineral at a Carolina beach is quartz, usually forming about 80–90 percent of the sand. Quartz is silicon dioxide (SiO_2) and is a very hard mineral, capable of rolling in the surf with little change. It has a clear, glassy appearance but is often stained light brown by iron.

The second most common mineral is feldspar. Feldspar is so similar in appearance to quartz grains that it is difficult to distinguish the two without a microscope. Although less durable than quartz, feldspar can still make up to 20 percent of the beach sand.

A third group of minerals in beach sand are the heavy minerals, or "heavies." These are discussed in Activity 13.

Did you ever wonder where sand comes from? The story on the next page narrates the journey of sand grains as they make their way down to the coast.

The Sands of Time

Africa and North America drifted toward each other for millions of years. Around 270 million years ago they finally collided. All the continents became united in the supercontinent, Pangaea. The collision crumpled the crust like a giant carpet, pushing up the Appalachian Mountains. Volcanoes formed, sometimes in violent eruptions, sometimes as lava spreading in slow-moving sheets.

But the supercontinent didn't last. About 230 million years ago, Pangaea began to break up. As the continents slowly drifted apart, the Atlantic Ocean was born. With no more mountain-building forces to offset erosion, the Appalachians eroded more quickly. Rivers transported sediments from the eroding mountains to the shore, forming the coastal plain.

One day, a powerful storm dislodged a boulder from high in the mountains. The boulder broke into smaller pieces and rolled into a creek. The creek became a river, and fragments from the boulder joined fragments from other rocks. Together they tumbled down the riverbed, becoming smaller and more rounded. Eventually, they became the size of sand grains. Millions of years later the grains were deposited on a sandbar where the river emptied into the sea.

Long years passed. Ice sheets covered the poles. Sea level dropped and the shoreline moved seaward. Then the climate warmed, the ice melted and the shoreline returned. Back and forth the shoreline moved as the climate alternated between periods of hot and cold. Sometimes the sand grains from the boulder were left on dry land, other times at the bottom of the sea or on a beach. Eventually, the shoreline arrived at its present location.

A child, gifted with that innate sense of curiosity all children possess, was playing at the water's edge. Pulling a magnet through the sand, the child held it up in the sunlight. Tiny black grains stood out from the edge of the magnet.

"Look, Daddy, magnetite! I wonder where it came from?"

As the story suggests, beaches in the Carolinas are made of sand grains, eroded from rocks and carried by rivers from the mountains (and also from the Piedmont) to the sea, a process that took millions of years. Even sand from the continental shelf that was put on a beach during storms ultimately originated in the mountains. That's the big picture. Now for the details.

A World in a Grain of Sand

WHAT YOU NEED A magnifying glass and beach sand.

WHAT TO DO Examine beach sand under a magnifying lens. Can you distinguish shell material from minerals? Why do sand grains have different colors? Do you see anything in the sand other than minerals or shells? Geologists define sand as sediment made of particles between $\frac{1}{16}$ millimeter and 2 millimeters (1 millimeter = .039 inches) in diameter, about the same size as the sugar in your kitchen. Use the chart below to classify the grain size of the sand at your beach.

Grain Size	Description
Boulders	Too big to carry
Cobbles	Bigger than grapefruit
Pebbles	Throwing size
Gravel	Peas
Very coarse sand	Lentils
Coarse sand	Coarse sugar
Medium sand	Granulated sugar
Fine sand	Superfine sugar
Very fine sand	Barely visible to the eye
Mud (silt and clay)	Not visible to the eye

More about Sand

Look closely at beach sand under a magnifying lens and you'll see a range of different colors. Each color is a different mineral or a shell fragment. Shell fragments can be distinguished from mineral grains by their irregular shapes, by their colors and by the fact that mineral grains are **better sorted** than shells. This means most of the mineral grains are about the same size.

Sometimes rounded and beautifully polished pebbles can be seen at the beach. These were deposited by ancient rivers that once flowed across the coastal plain when sea level was lower. The beach you are standing on now was once part of the mainland with rivers streaming toward a distant shore, miles to the east.

Depending on your location on the beach you might also see organic material in the sand—anything from driftwood and marine algae to sea life tossed ashore by wind and water. The organic component is an important food source for crabs and shorebirds.

There is another component of beach sand, one that didn't exist until modern times. Plastic, glass, steel and other discards from industry are showing up on the world's beaches. Most plastics don't decompose. Instead, they are broken up by waves into ever-smaller particles, which eventually become part of the beach sand. Problems with plastic are discussed in Activity 33.

Rounded pebbles can be found on many Carolina beaches.

Black Sand

We take a handful of sand from the endless landscape of awareness around us and call that handful the world.

—from Robert Pirsig, *Zen and the Art of Motorcycle Maintenance*

Often mistaken for an oil spill, the patches of black sand so common on the world's beaches range in width from a few inches to an entire beach blackened from top to bottom. Usually the black sand is concentrated on the upper beach. But if you look closely at sand from anywhere on the beach you can almost always spot a few black grains.

Black sand is also called heavy mineral sand. To find out why, try the following simple experiment: Locate a thick patch of black sand. Grab a handful of the sand. Next pick up an equal amount of beach sand that has no black grains. Which handful weighs more?

Black sand is clearly heavier than sand with few or no black grains. The heavy minerals, or heavies, that make up black sand ultimately originated in the Appalachians and Piedmont. Rivers carry heavy minerals (as well as light minerals like quartz) from distant sources to the sea. Only the carbonate fraction (shell part) of beach sediment is formed near the beach.

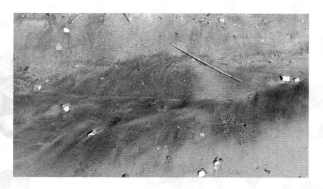

Black sand on the
Outer Banks, N.C.

Identifying Minerals

Black sand consists of a variety of different minerals, many of which, if they were just a bit larger, would be considered semiprecious gemstones. This activity will help you identify some of these minerals.

WHAT YOU NEED Black sand, a magnifying glass and a magnet.

WHAT TO DO Drag a magnet through a patch of black sand (best done with dry sand). The grains picked up by the magnet are mostly magnetite with smaller amounts of ilmenite. Both of these minerals are magnetic, though ilmenite is less so than magnetite. After magnetite and ilmenite have been removed from the sand you can identify the other heavy minerals by their colors. Examine the remaining grains under a magnifying lens and note their colors. Ideally, this should be done on a bright, sunny day. You can also do this activity at home with a microscope under low magnification. Using the chart below, how many minerals can you identify?

Mineral	Color
Light minerals	
Mica	silver, black
Quartz	clear, gray, white, brown, yellow
Heavy minerals	
Amphiboles	dark green, black
Epidote	yellow-green, green
Garnet	red, pink, purple
Ilmenite	black
Kyanite	gray, blue
Magnetite	black
Rutile	red, black
Topaz	clear, gray, yellow, blue
Tourmaline	black, green, pink
Zircon	colorless, brown

Rare minerals such as topaz or zircon require a powerful microscope and considerable expertise for positive identification. Other minerals like garnet or epidote are easier to recognize. As a first guess without a sophisticated microscope, pink or purple grains are garnets, red may be rutile, black and green could be tourmalines, yellow-green are epidote and dark green minerals are probably amphiboles (like hornblende) or pyroxenes.

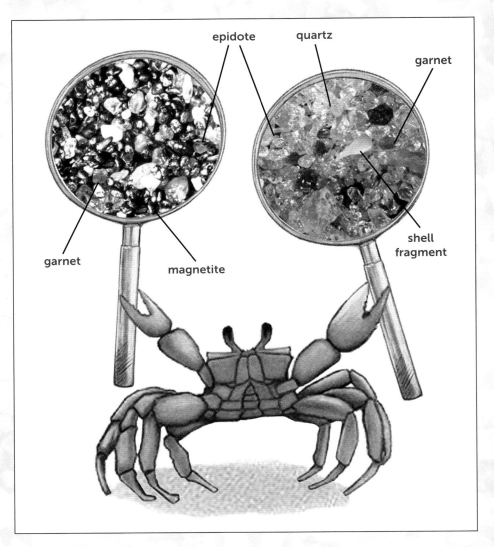

epidote

quartz

garnet

garnet

magnetite

shell fragment

Garnet-rich sand on
Bogue Banks, N.C.

More about Heavy Minerals

Why do heavy minerals sometimes get concentrated into spectacular patches or layers? Less heavy grains like quartz are more easily carried away by wind or water, leaving heavier, darker minerals behind.

Why are heavy minerals considered to be beach fingerprints? Each of the rivers that bring sand to the sea transports different groups of heavy minerals. By analyzing heavy minerals in beach sand and comparing them to those in river sands, you can name the river that brought the sand to the beach.

Are semiprecious heavy minerals ever big enough or plentiful enough to be mined? Beaches are mined worldwide, usually to get sand for concrete or for **beach replenishment** projects, but also to extract gold, diamonds, tin, staurolite and magnetite. Beach mining is environmentally bad; it erodes beaches and harms the plants and animals living there.

DID YOU KNOW?

Scientists think magnetite in the bodies of birds, sharks, sea turtles and other critters helps them navigate by allowing them to align with the earth's magnetic field. Think of magnetite as nature's GPS!

Barking Sand

Even by day men hear these voices of spirits . . .
—Marco Polo (thirteenth-century Italian merchant
who wrote about his travels to China)

It's called barking sand, squeaking sand, singing sand, musical sand or in Japan, frog-sound sand. It is found on almost all of the world's beaches. Why does sand sometimes make sounds?

WHAT YOU NEED Patience and energy to shuffle around on the beach.

WHAT TO DO Walk along the upper dry beach above the high-tide line. Shuffle your feet as you walk and listen for a sound made by the sand. The sound is somewhere between a bird's chirp and a small dog's bark. If you've been around beaches a lot, you've probably heard the sound before but may not have noticed it. As you are shuffling along, some people hanging out on the beach may stare and wonder what you are doing. Don't worry about these curious strangers. They are lazing around while you're working hard at probing the mysteries of nature! After you find barking sand, consider the following questions:

> What is the nature of the grains that produce barking sounds? Are they large or small, with lots of shells or few shells?
> Where do patches of barking sand occur? How big are the patches?
> What is the character of the barking sand? Is the sand wet or dry?

Why Sand Barks

Actually, there have been few scientific studies of barking sand on beaches. This means you are doing original research and your theories and observations may add to our knowledge of beach sand. Some think the barking or singing sound is due to layers of sand grains sliding over one another. Others believe the sound is produced by cushions of air in the sand or even by encrusted salt coating individual sand grains.

We have observed that barking usually (but not always) occurs in dry, well-sorted sand without shell fragments and with lots of quartz. Well-sorted sand means that the grains are all about the same size.

Not all sands bark alike. Small grains (fine sand where the grains are barely visible to the naked eye) produce weak sounds. Medium sand grains (the size of sugar particles) produce a wide variety of sounds, often quite loud. Sand with mud or sand that is polluted makes no sounds.

Dune sand in many of the world's deserts also makes sounds. The sounds, usually described as deep booms, are caused by sand slipping down a dune face. In the thirteenth century Marco Polo heard booming sands while crossing the Gobi Desert. In his mind he was hearing the voices of evil spirits.

EXPLORE YOUR WORLD

Experiment by adding a small amount of water to sand on the beach and see what happens to the sound. Record barking sand on your digital device or listen to the many recordings of sand sounds posted on YouTube.

Soft Sand

I was exceedingly surprised with the print of a man's naked foot
on the shore, which was very plain to be seen in the sand.
—from Daniel Defoe, *Robinson Crusoe*

Have you ever been walking along a beach, the sand pleasantly firm beneath your feet, when suddenly you sank ankle deep into the sand? What happened? Chances are you found a patch of soft sand.

Most beaches have scattered areas of soft sand. In some rare cases, such as on Hilton Head and other nearby beaches, the sand may be so soft you will sink up to your shinbone. This activity explores the phenomenon of soft sand, where and why it occurs and why it's important.

Looking at Soft Sand

WHAT YOU NEED Curiosity and patience.

WHAT TO DO Walk along the beach until you find a stretch of soft sand. Make sure to walk both the lower beach and the upper beach, including by the dunes. Is soft sand on the low tide beach, the high tide beach, or above the high tide beach? Is the soft sand shelly (like a **shell hash**) or are there few shells present?

After finding a patch of soft sand, force your hand into the sand and extract a chunk of the sand with a fresh face on it. This may take some practice (you can also use a shovel). Can you see the bubbles (circular holes) in the sand? Now you understand why geologists call soft sand **bubbly sand.** The bubbles consist of air pockets. When you step on a patch of bubbly sand, the bubbles collapse and your foot sinks into the sand.

You probably observed that soft sand occurs in long stretches on a beach near the high tide mark but rarely at the low tide mark. Why is this so? If you examine the tidal chart for Activity 6 you will see that greater **tidal ranges** (the distance between high and low tide) occur the farther south you go. Curiously, deeper layers of bubbly sand are often found in South Carolina where the tidal amplitudes are relatively high. Is there a connection between tides and the formation of soft sand? Any idea what that connection might be?

Bubbly sand can be an annoyance to people walking, jogging or driving on a beach. But as we shall see on the next page, it also has serious environmental consequences for beaches, especially where an oil spill has occurred.

How Soft Sand Forms

Bubbly sand forms as the result of air being trapped in the sand as the tide is rising. When the tide goes down, air replaces seawater in the spaces between the sand grains. When the tide comes back, seawater forces the air upward. Most of the air escapes back to the atmosphere, but some air gets trapped under layers of wet sand. In the right kind of sand (fine-grained sand with few shells), air pockets form and the sand becomes bubbly.

The bubbly sand layer is deeper on Hilton Head Island because the tidal range is higher there. This allows a larger amount of air to get into the sand. The incoming tide forces this air upward, creating a thicker band of bubbles. This also explains why soft sand rarely occurs on the low tide beach. At low tide the water isn't able to push as much air into the sand. Bubbly sand is usually better developed in well-sorted, fine-grain sand with few shells because coarser, shelly sand allows more air to escape.

The presence of soft sand has an important environmental implication. Because of its many air pockets, soft sand allows oil from an oil spill to penetrate deeper into the sand than into other sands. Liquid pollutants (oil, sewage, chemical spills) can easily flow from one air pocket to another. This makes cleanup harder. By contrast, a beach with firm sand and few bubbles allows little penetration of pollutants, making cleanup easier.

4

Beach Features

There are more things in heaven and earth,
Horatio, than are dreamt of in your philosophy.
—from Shakespeare, *Hamlet*

Beach features are the bubbles, burrows, tracks, trails, pits, pedestals, ripples and countless other traces on sand left by wind, water or life. Some have been extensively researched by scientists. Others have yet to be discovered. Most are fleeting patterns that appear or disappear in minutes or seconds, according to the whims of wind and tide. Understanding these features requires recognition followed by a little deductive thinking to figure out how they were formed.

The most common and easily recognizable beach features are ripple marks. Ripples are of great interest to geologists. Not only do they reveal recent events on a beach, but when preserved in sedimentary rocks they can tell us about events that happened millions of years ago. Other common features are holes formed by the interaction of air and water and holes made by burrowing animals. Both marine and land animals also leave their tracks and trails on the surface of the beach, and groundwater creates interesting patterns as it leaks from under the sand at low tide.

Not every beach feature is on the surface. To see the layers of sediment (sand) that make up a beach you must dig a ditch with a shovel.

Science, it would seem, requires both physical and mental energy.

Ripple Marks

If you do not change direction you may end up where you're going.
—Lao Tzu (Chinese philosopher and mystic)

Ripples are small ridges on the beach surface and on dunes. They form when either waves, water currents or winds transport sand. Ripples can be quite variable and complex. Most are variations of two basic types: symmetrical **wave ripples** and asymmetrical **current ripples**.

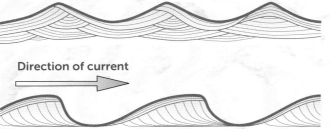

Side views of ripples. Wave ripples are made by waves. Current ripples are made by water currents or air currents. Their shape indicates the direction of the current flow.

WAVE RIPPLES

These are symmetrical ripples where both slopes are identical. Wave ripples are formed by the up-and-down movement of waves where little forward motion of the water is present.

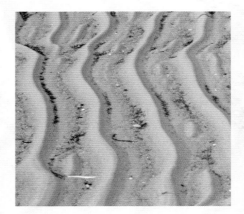

CURRENT RIPPLES

These are usually the most common types on both beaches and dunes. In a side view each ripple has two slopes. The slope facing the current is gentle, while the slope facing away from the current is steep. By looking at the shape of the ripples you can tell the direction of the current that formed them.

DUNE RIPPLES

Wind makes these ripples. They are current ripples but are usually smaller than the current ripples on the beach and are made of finer sand grains. What makes dune ripples unique is that they may be reoriented in a few hours if the wind changes direction.

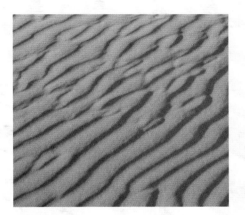

LADDERBACK RIPPLES

Ladderback ripples are formed when there are two currents coming from different directions, one after another for brief intervals. The resulting ripples are a ladderlike pattern. They often form with the coming and going of swash.

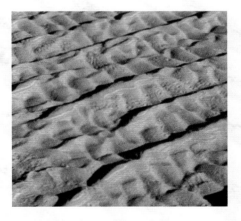

ADHESION RIPPLES

Usually seen above the high-tide line, these special ripple marks don't look like other ripples. They form when strong winds coincide with rain.

How many kinds of ripples can you spot at your beach? Are there beach features other than ripples made by wind and water?

WHAT YOU NEED A notebook, a pencil and curiosity.

WHAT TO DO Wander around the beach. Sketch the ripples you see and note their locations. Do this activity on different days and note any changes. If you're lucky, you may be able to watch the ripples form. On a windy day you may see ripples migrating on dune surfaces. Take notes on what you see and explain the ripple formation processes. The examples in this activity are but a small sampling of the countless forms found at a beach. In addition to ripples, there are other features made by wind and water. Look at the photos on the next page. Find similar features and explain their origin. And if you encounter a mysterious feature like the one pictured here that defies explanation, take comfort in the knowledge that nature is more profound than our theories.

Beach grass and wind traced this near-perfect circle.

Pedestals are made by wind erosion when underlying layers of sand are less resistant to erosion than the top layers.

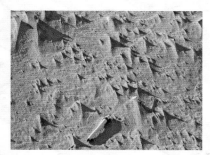

Wind shadows behind shells allow sand to accumulate. Which way is the wind blowing?

When saltwater in beach sand evaporates, it leaves behind salt, which cements the grains together, creating thin crusts of sand called salcrete.

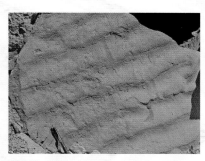

Fossil ripples.

DID YOU KNOW?

Fossil current ripples can tell geologists current direction on ancient beaches. Can you guess which way the current was flowing in the photo above?

How do you think these circles formed?*

*Answer: raindrops

Volcanoes on the Beach

What has been will be again, what has been done
will be done again; there is nothing new under the sun.
—Ecclesiastes 1:9

Try the following experiment. Walk along a beach when the rising tide has almost reached the high tide mark. Look carefully on the wet sand for holes that form as the waves recede. You should see tiny holes the size of small nail holes forming in the sand. Some of the holes may be in the centers of small pits. A few of the pits may have raised sides that look like tiny volcanoes. Each beach will be a little different, but look long enough and you will find black and/or white rings of sand that are an inch or more in diameter. And on some beaches you will find raised mounds of sand that resemble blisters on the beach surface. Nail holes, pits, volcanoes, rings and blisters are features common to sandy beaches all over the world. They appear, disappear and reappear according to the rhythm of the tides. They have been doing so for as long as there have been tides and seas and sand.

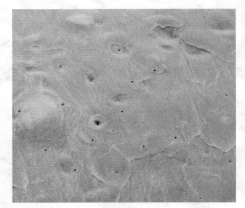

Holes, pits and blisters.

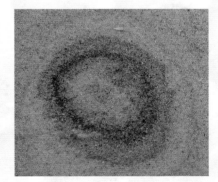

Ring (2 in. dia.). Myrtle Beach, S.C.

WHAT YOU NEED A beach just before high tide, a pocketknife, a notebook, a pencil and an inquiring mind.

WHAT TO DO Look for holes and any of the features mentioned on the previous page. When you think you've located holes, pits, rings or blisters, wade about ankle-deep into the swash. Observe the water as it rushes up to you and watch for bubbles streaming out of the sand into the water. The bubbles coming up from the sand are from air trapped between sand grains. This air is the mechanism by which the holes are created. (To learn how air gets into beach sand, see Activity 15.)

When the tide has noticeably gone down, get on your hands and knees and carefully examine the surface of the beach for holes made from escaping air. You may have to wander around to find well-developed holes. Note the different types of holes and draw them in your notebook. Also note and describe any rings you encounter.

If you find a blister on the beach, try the following: With a pocketknife carefully cut the blister in half and scrape away enough sand to see inside. What do you see? Can you guess how blisters form? What about rings, pits or volcanoes? In your notebook write a few sentences outlining your hypothesis (an educated guess; see Activity 3) explaining how rings, volcanoes, pits and blisters are made.

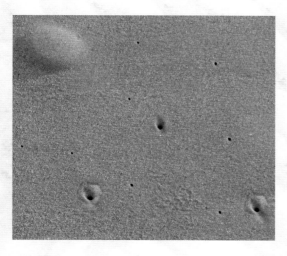

Blister (2 in. dia.) and different kinds of holes. Atlantic Beach, N.C.

Holes in the Sand

BLISTERS

As the tide rises, seawater seeps into the beach and forces air out of the sand. For reasons not always understood, some air escapes and some gets trapped in the sand. Seawater pushes the trapped air upward. The sand expands like a balloon and forms a blister. If you slice a blister in half and scrape away the excess sand you will discover an air pocket in the center.

Side view of a blister.
(Coin for scale.)

NAIL HOLES

These are by far the most common type of holes formed by escaped air. They are never larger in diameter than a small nail.

VOLCANOES

When sand gets forcibly ejected by air, it piles up and forms craters that resemble miniature volcanoes. Sand volcanoes are also made by groundwater bubbling out of the beach at low tide.

PITS

Pits are likely made by several different ways. Some are created like volcanoes, but there is less air to force the sand into a cone. Some are made when waves wash away a blister, leaving a depression. Others are made by groundwater at low tide.

RINGS

Sometimes blisters are formed in sand that has a layer or layers of dark, heavy minerals. A wave washing over the sand removes the top of the blister, leaving a ring of heavy minerals. Rings can be black or white and sometimes create complex patterns (see below).

Make Your Own Holes

WHAT YOU NEED A beach just before high tide and a sense of fun.

WHAT TO DO Place some driftwood in the swash where waves will wash over it. If conditions are right, a series of holes will form outlining the driftwood as the wave recedes. Use your imagination to find ways to enhance the formation of holes in the sand. Can you create your own volcanoes?

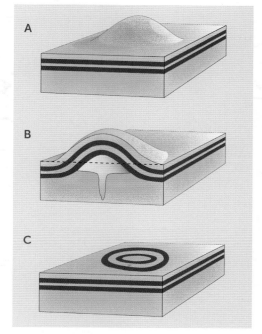

How rings are formed
A. Blister in sand with layers of heavy minerals.
B. Side view of blister.
C. Rings formed after a wave washes away blister.

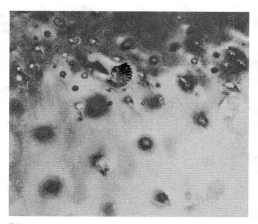

Rings

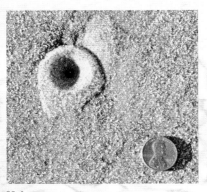

Volcano.

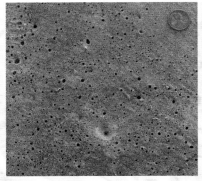

Pinholes and pits.

Air escaping from sand through seawater.

Blister.

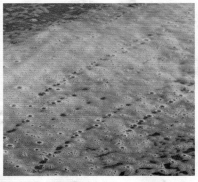

Line of bubbles in swash from which holes will form.

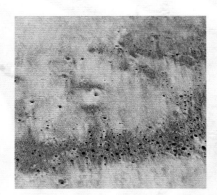

Pits and nail holes in black sand.

Beach Tracker

*If you talk to the animals they will talk with you and you will know
each other. If you do not talk to them you will not know them and what
you do not know, you will fear. What one fears, one destroys.*
—Chief Dan George (Native American actor and writer)

At first glance a beach appears to be an uninhabited stretch of sterile sand, as empty as an old shell. In fact, a natural beach teems with life (though we often fail to notice it) and the land between the forest and the sea has many uses for animals—as a home, a rest area, a navigation aid, a nesting site and even a fast-food diner!

A beach is also a place where land animals and the coastal ecosystem sometimes mix. All over the world animals from the land prowl the shoreline and leave their tracks on beaches—leopards in South Africa, kangaroos in Australia, tigers in Siberia and elephants in Ghana. We have not yet spotted any of these animals on a Carolina beach, but we have seen tracks left by some other critters, including raccoons, bobcats, turtles, coyotes, crabs, horses and dogs.

This activity is about the tracks, trails, burrows and other animal signs common to Carolina beaches. Tracks are the remains of an animal's passage (usually its footprints) across the beach. Trails are the traces that a burrowing animal like a snail makes as it crawls under the surface of the beach. Burrows are the underground homes of animals that live in the sand.

Tracks, Trails and Burrows

Like any kind of tracking, this activity requires patience and persistence. Trails are not particularly abundant, and you should expect to do a lot of walking. Most of the animals that make trails in the sand are snails. The following is a list of some common snails that leave trails on Carolina beaches (see Activity 21 for images):

Shark eye
White baby ear
Lettered olive
Coquina

WHAT YOU NEED Energy for hours of wandering on a beach, a notebook and a garden trowel or tablespoon. A book or pamphlet on Carolina shells would be useful.

WHAT TO DO Best carried out on a beach that has not been replenished and isn't jam-packed with bathers whose footprints would destroy the trails. Low tide would be the best time. Wander along the beach looking for trails, tracks and burrows. When you find a trail or a burrow, get down on your hands and knees and carefully dig up the sand to identify the responsible critter. *Do not harm* the animal. Identify any tracks. Record your discoveries in sketches or photographs. Do this at different tide levels on your beach, and if possible repeat the experiment on other beaches. If you come back in another season or after a storm or after a replenishment project, do the same thing.

WHAT TO LOOK FOR

Look for lines, holes or mounds of sand that may indicate a critter's home. Check both the wet and dry beach.

Lugworm (*Arenicola marina*) mounds. Lugworms look like earthworms. They live in U-shaped burrows, eating sand that goes through their intestines and forms spaghettilike mounds. Birds and fish eat lugworms.

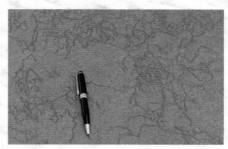

Worm trails.

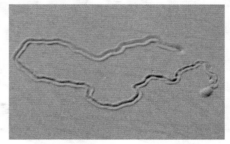

Snail trail.

Did you know?
We ghost crabs dig our burrows above the high-tide line but must return to the sea several times a day to wet our gills in order to breathe.

Ghost crab tracks and burrow. Ghost crabs eat mole crabs, clams and turtle eggs. In return they are food for gulls and raccoons.

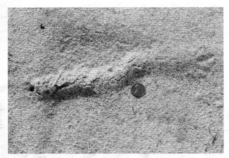

Mole trail on the upper beach just below the dunes.

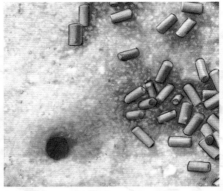

Carolinian ghost shrimp (*Callichirus major*) burrow surrounded by fecal pellets. Ghost shrimp burrow in the intertidal zone. Their poop (pellets) is eaten by hermit crabs and other scavengers.

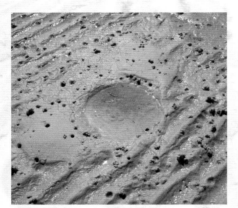

Resting place for a ray.

You may occasionally find snake tracks in the grasslands behind the dunes but almost never on the beach. Nevertheless, the copperhead in this picture was found swimming in the surf on Shackleford Banks. Notice the black-stained shells around the snake.

Loggerhead turtle tracks.

WHO WALKS ON THE BEACH?

Insects, spiders, lizards, an occasional snake, raccoons, coyotes, foxes, bobcats, mink, wild boar, deer (rarely), birds, mice, moles, turtles, crabs, worms, mollusks and many other critters leave tracks on beach sand.

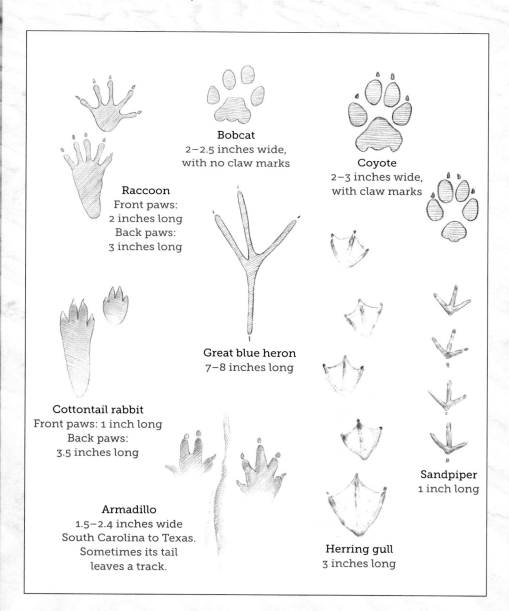

Raccoon
Front paws:
2 inches long
Back paws:
3 inches long

Bobcat
2–2.5 inches wide,
with no claw marks

Coyote
2–3 inches wide,
with claw marks

Great blue heron
7–8 inches long

Cottontail rabbit
Front paws: 1 inch long
Back paws:
3.5 inches long

Armadillo
1.5–2.4 inches wide
South Carolina to Texas.
Sometimes its tail
leaves a track.

Herring gull
3 inches long

Sandpiper
1 inch long

SKETCHBOOK JOURNAL (NATURE JOURNAL)

A sketchbook journal is part diary, part sketchbook. It includes a person's thoughts, ideas, observations and experiences. Many famous explorers, scientists and artists, such as Lewis and Clark, Charles Darwin and Leonardo da Vinci, kept sketchbook journals. You should do the same. Keep a sketchbook journal of your scientific explorations of the beach. Record any trails or tracks you come upon. Draw (with pencils, colored pencils, pens, etc.) interesting things you find on the beach. Try to keep the same journal going for several years, recording what you see and do each time you visit the beach.

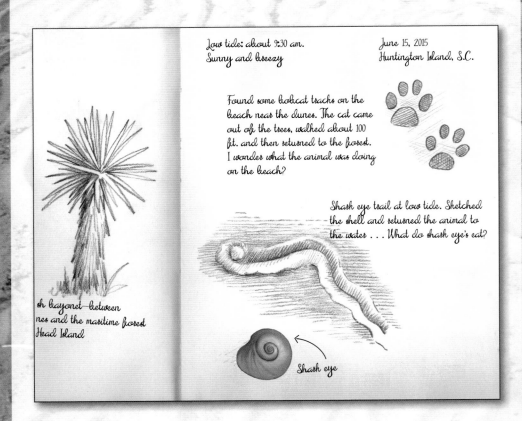

Low tide: about 9:30 am.
Sunny and breezy

June 15, 2015
Huntington Island, S.C.

Found some bobcat tracks on the beach near the dunes. The cat came out of the trees, walked about 100 ft. and then returned to the forest. I wonder what the animal was doing on the beach?

Shark eye trail at low tide. Sketched the shell and returned the animal to the water . . . What do shark eye's eat?

Spanish bayonet—between dunes and the maritime forest Hunting Island

Shark eye

The Layer Cake Beach

We are tied to the ocean. And when we go back to the sea,
whether it is to sail or to watch—we are going back from whence we came.
—President John F. Kennedy

Beaches are made up of layers of sand, not unlike the layers in a cake, except the layers in a beach are usually very thin (and not so tasty). Each layer represents an event, sometimes large, sometimes small. The event may have been a blast of wind, a wave swash, a storm wave or some combination of events.

To observe beach layering it is first necessary to dig a ditch that (and we can't emphasize this enough) should *not* be more than one foot deep. Every year children are hurt when the sides of a hole excavated on a beach collapses on them.

WHAT YOU NEED A small shovel and the energy to dig a ditch.

WHAT TO DO Dig a one-foot-deep ditch down the beach slope, perpendicular to the shoreline. It should be a shovel width wide or more and 10 feet or so long. Put all the excavated sand on one side of the ditch. Now carefully smooth off the surface of the other side of the ditch so you can see the layers. Keep your friends away from the edge of the ditch because it will inevitably collapse and have to be repaired. When you finish observing the layers, fill in the ditch. If you are feeling ambitious, try digging a new ditch at a different location or at the same location after a storm.

Riddles in the Sand

What are the different layers composed of?

Which layers were caused by wind depositing sand and which by water (waves and currents)?

Most ditches will show layers of shells or shell fragments. Can you guess how shell layers were formed?

Cross section of a beach in South Carolina showing black layers of heavy minerals. Note the raccoon tracks at the top. Humans are not the only critters doing beach activities.

The Layer Cake Beach

The answer to these questions is difficult because there are so many things happening at the same time on beaches. You might expect wind layers to be thin, perhaps just a few sand grains thick. The thicker layers of shell fragments may have been deposited by big waves or may represent a quiet period when the wind took away the small sand grains, leaving behind a layer of shells. The dark layers are from heavy minerals (see Activity 13). On beaches with lots of people you may see buried footprints in the form of distorted layers, and you may see layers distorted with organic matter (plant material, driftwood, etc.).

The important discovery you've made is that beaches are made of layers of sand and shells that formed under a variety of conditions. To see how waves and currents form those layers, do the following activity.

Make Your Own Layers

WHAT YOU NEED A jar (15 ounces or bigger), seawater and some beach sand with heavy (black) minerals and small shell fragments.

WHAT TO DO Fill the jar about one-third full with sand and fill the rest of the jar with seawater. Make sure your sand has a lot of heavy minerals (black sand) as well as small shell fragments. Now turn the jar upside down and shake it vigorously for 20 or 30 seconds. Put the jar right side up on a flat surface and wait for everything to settle. Experiment with different ways of shaking the jar and find a technique that creates the best layering. Now can you understand how layers of sediment are created at the beach? The sand and shells that are suspended (floating) in the water begin to sink when the water stops moving. The heavier (denser) black sand drops first, the less dense light-colored sand drops next and the pieces of shells drop last, thus forming layers.

With a little practice and vigorous shaking of your jar you can make layers of sand similar to the layers in beach sand.

REFRIGERATE

Groundwater

In one drop of water are found all the secrets of all the oceans.
—Kahlil Gibran (Lebanese American poet)

People go to beaches for many reasons. Some to collect shells. Some to swim. Others to fish or to laze in the warm sunshine. What few realize is that their recreational pursuits are accomplished over an enormous "pool" of underground water just a few feet under the sand.*

Water found below the surface of the earth is called groundwater. What's unique about groundwater at the beach is that you don't need to dig a hole to find it. Stand near the surf at low tide and you can watch groundwater seeping out of the beach and into the sea. Sometimes the groundwater carves patterns in the sand known as rill marks. Rill marks (or rills) look like miniature rivers. The goal of this activity is to demonstrate the presence of groundwater at the beach, to determine whether that water is fresh or salty and to learn why groundwater is important to the beach ecosystem.

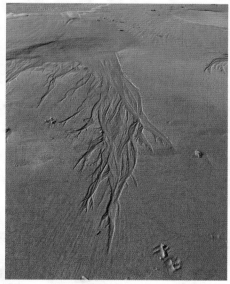

Rills sometimes appear where groundwater flows out of a beach.

*The water under the sand of course is not literally a "pool" but is rather a large mass of water between grains of sand.

Finding Groundwater

WHAT YOU NEED A beach at low tide and a hydrometer. A hydrometer is a device that measures salinity (how much salt is in water). Hydrometers can be found at most pet stores.

A hydrometer

WHAT TO DO Walk along the lower beach at low tide and look for signs of groundwater outflow. This could include rill marks or wet sand patches above the low-tide line. When you think you've located groundwater, dig a hole deep enough to fill the hydrometer with water and measure the salinity. *Do not* taste the water. It may be contaminated with bacteria.

Hydrometers measure salinity in parts per thousand. Tap water should have a hydrometer reading close to zero. Seawater should be in the low 30s (30–36 parts per thousand [ppt]). After measuring groundwater salinity, do the same for the nearby ocean water. For this experiment classify salinity as follows:

Hydrometer reading	0–5 ppt	5–30 ppt	Above 30 ppt
Salinity	Fresh	Brackish	Salty

Sand That's Full of Water

Sand under a beach is saturated by a lens of freshwater "floating" on a denser layer of saltwater. The freshwater comes from rain. The saltwater comes from the ocean. On barrier islands, freshwater slowly flows into the sea, both from the sound side and from the ocean side.

When the tide goes down, the freshwater moves outward into the sea and mixes with saltwater, which is also flowing outward. When the tide comes in, the freshwater is pushed back into the beach sand. The groundwater you see at the beach will most likely be brackish or salty, except after a heavy rain, when it may be fresh.

Pizza Delivery

As groundwater creeps through the beach it transports nutrients, such as nitrogen, that support algae, fungi and bacteria living in the sand. These microorganisms are food for larger meiofauna (critters 0.5–1 mm in size), which in turn are food for mollusks, crabs and fish. Groundwater is like a pizza delivery car bringing delicious snacks to a hungry beach community.

Not only does groundwater deliver nutrients for critters in the sand, but it also supplies freshwater for thirsty plants and animals behind the dunes, especially in low-lying marshy areas known as swales. Swales are where the water table (the upper surface of groundwater) is exposed. For thirsty raccoons, rabbits, bobcats, birds and other life on barrier islands, the groundwater exposed at a swale is a welcome source of drinking water.

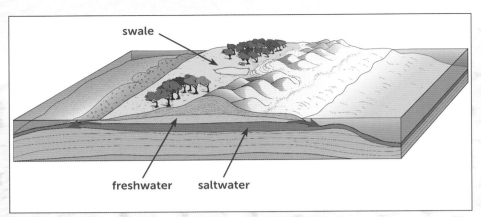

swale

freshwater   saltwater

Cross section of a barrier island showing sand saturated with a layer of groundwater (freshwater) on top of a layer of saltwater. Red arrows indicate where groundwater is flowing into the ocean.

Sometimes an entire beach is overflowing with groundwater, like this one on Edisto Island. Rills may or may not be present. Note the groins extending seaward across the beach.

CHAPTER

5

Shells

I have a large shell collection which I keep scattered on beaches all over the world. Maybe you've seen it.

—Steven Wright (comedian, actor and film producer)

People have been collecting shells as long as there have been people living near the sea. Native Americans used them for tools, money and jewelry. Some tribes wove shells into belts called wampum, which they used to help them remember tribal history. Others used them for musical instruments. Today people collect shells for their beauty, but they seldom stop to consider what shells are and what roles they play in the beach ecosystem.

Shells or, more correctly, the animals that live in shells are an essential part of a complex food web that includes fish, birds, mammals, humans and a score of other marine and land animals. But what do the animals in the shells eat? What happens to a shell after its occupant dies? Why are some shells so brilliantly colorful while others are black or brown, like rusted steel? How many years have the seashells you see on the beach been lying there? What are shells made of? What other interesting collectibles besides seashells wash up on a Carolina beach?

ACTIVITY 21

Shells

One cannot collect all the beautiful shells on the beach;
one can collect only a few, and they are more beautiful if they are few.
—from Anne Morrow Lindbergh, *Gift from the Sea*

Seashells are the protective outer bodies of **invertebrates** (animals without backbones) that live in the sea. Shells are made of layers of calcium carbonate ($CaCO_3$). As the animal grows, new layers of calcium carbonate are added and the shell expands. When the animal dies, the softer parts of its body rot away, leaving only the shell.

Few realize that many of the shells that wash up on American beaches are in fact fossils. They may be colorful and shiny, looking as if the organism that made the shell died yesterday, but they are hundreds or even thousands of years old.

Collecting Shells

Seashells come from three main groups: **mollusks**, **echinoderms** and **crustaceans**. Mollusks, the most common, are of two types: 1) **gastropods** (or snails), which have one shell, and 2) **bivalves** (clams, scallops and oysters), which have two shells. Cephalopods (squids and octopi) are also mollusks, but they rarely have external shells. Sea urchins and sea stars are echinoderms, while crabs and barnacles are crustaceans.

WHAT YOU NEED Ziploc bags, a shell field guide and some restraint.

WHAT TO DO It's best to look for shells at low tide. Winter is the preferred season because there are fewer beachcombers and there are frequent storms that wash shells up on the beach. As you add to your collection learn about the roles the organisms play in the marine ecosystem. What do they eat? Who eats them?

Etiquette for Shell Collectors

Don't be greedy. Leave some shells for the next collector. Specimen collection is an important tradition of science, *but* our beaches have become stripped of shells because so many people collect them. Instead of collecting 20 or 30 shells, we recommend taking one or two quality specimens for your collection. *Or* start a shell photograph collection instead. *Or* return the shells to the beach after you've studied them.

Leave the beach as undisturbed as you found it. This means don't litter, clean up any litter you find (see Activity 33) and don't trample dune plants.

Never take a living organism. This includes living shells on the beach as well as plants and animals behind the dunes.

GASTROPODS (SNAILS)

Most gastropods are carnivores or scavengers. Many eat other mollusks, including members of their own species. Birds, crabs, fish and other critters also eat gastropods.

Lettered olive (*Oliva sayana*).
2.5 inches long.
Food: coquina and other clams.
South Carolina state shell.

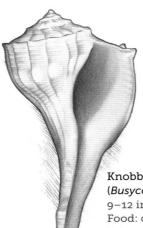

Knobbed whelk
(*Busycon carica*).
9–12 inches long.
Food: clams and oysters.
Pries open clams by using
the edge of its shell.
Migrates to deep water
during storms.

Giant eastern murex
(*Hexaplex fulvescens*).
7 inches long. Food: oysters
and other bivalves.

Shark eye or moon snail
(*Neverita duplicata*).
3 inches long.
Food: clams and other
mollusks.

Slipper snails (*Crepidula* spp.)
1 inch long. Slipper snails
don't feed like other snails.
Like clams, they filter food
from seawater instead.

Scotch bonnet
(*Semicassis granulata*).
4 inches long. Food: sea
urchins and sand dollars.
Predator: blue crabs.
North Carolina state shell.

Baby ears (*Sinum* spp.).
3 inches long.
Food: bivalves.

Florida horse conch
(*Triplofusus giganteus*).
19 inches long.
Food: clams, snails and other
horse conches.
North America's biggest snail.

Atlantic oyster drill
(*Urossalpinx cinerea*).
1.5 inches long.
Food: oysters, barnacles,
mussels.
Secretes acid to weaken
oyster shell, then drills
a hole and eats the soft
innards of oyster.
Hole tapers to a point.

Thick-lipped drill (*Eupleura caudata*).
1.5 inches long. Food: oysters.

Florida rocksnail
(*Stramonita haemastoma*).
3 inches long.
Food: barnacles, clams and
gastropods.
Found on rocks and jetties.

Banded tulip   True tulip

Tulip snails (*Fasciolaria* spp.).
4 inches long.
Food: snails and other tulip snails.
Banded tulips are smaller than true tulips.

Marsh periwinkle
(*Littoraria irrorata*).
1 inch long.
Food: fungi in salt
marshes.

Greedy dovesnail
(*Costoanachis avara*).
.5 inches long. 12 ribs
on body. Scavenger
and carnivore.

Giant tun (*Tonna galea*).
10 inches long.
Food: mollusks, sea
cucumbers and fish.

Atlantic deer cowrie
(*Macrocypraea cervus*).
5 inches long. Food: algae.

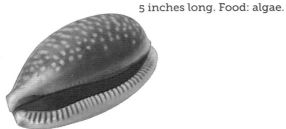

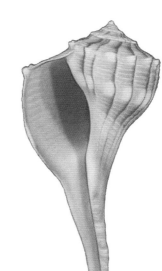

Lightning whelk
(*Busycon sinistrum*).
16 inches long.
Food: bivalves.
Unlike most gastropods, which
are right-handed, the lightning
whelk is left-handed (its cavity
is on the left side).

Clench helmet (*Cassis madagascariensis*).
12 inches long. Food: sea urchins.

BIVALVES (CLAMS, OYSTERS, MUSSELS, ETC.)

Most bivalves are **filter feeders** and feed on **plankton**. Their gills draw plankton (and oxygen) from seawater. Some bivalves bury themselves in sand and filter plankton with a straw-like siphon.

Atlantic calico scallop (*Argopecten irradians*). 2–3 inches long. Predators: gulls, squid, sea stars and crabs.

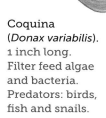

Coquina (*Donax variabilis*). 1 inch long. Filter feed algae and bacteria. Predators: birds, fish and snails.

Northern quahog (*Mercenaria mercenaria*). 4 inches long. Food: algae and diatoms. May live as long as 40 years.

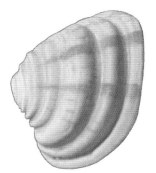

Imperial Venus clam (*Chione latilirata*). 1.4 inches long. Predators: horse conchs, shark eyes and other gastropods.

Jingle shell (*Anomia simplex*). 2 inches long. Filter feed.

Jackknife clam (*Ensis megistus*).
5 inches long.
Burrows deep in offshore shoals.
Predators: gulls, oystercatchers,
worms and shark eyes.

Stout tagelus (*Tagelus plebeius*).
4 inches long. Grows in salt marsh.
Found on beach after barrier island
migration (see Activity 11).
Predators: oystercatchers.

Cut-ribbed ark
(*Anadara floridana*).
4.5 inches long.
Predators: various
gastropods.

Penshells (*Atrina* spp.).
9–11 inches long.
Large, delicate shells with a
mother-of-pearl luster on the
interior of the shell.
Predators: starfish, conchs and
other gastropods.

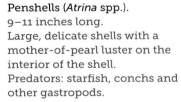

Angelwing (*Cyrtopleura costata*).
7 inches long. Burrows deep into clay or
mud offshore and siphon feeds.
Predators: fish.

DID YOU KNOW?

Filter feeders like oysters and mussels clean seawater of bacteria and toxins.

Pricklecockles (*Trachycardium* spp.**).**
2–3 inches long. Pricklecockles can jump several inches to escape predators like sea stars and ducks.

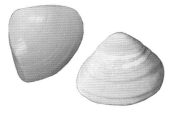

Keyhole limpet (*Diodora cayenensis***).**
2 inches long.
Attaches to rocks on groins and jetties.
Predators: sea stars.

Blue mussel (*Mytilus edulis***).**
2.5 inches long.
Attaches itself to rocks, piers, groins and jetties.
Predators: gulls, sea stars and humans.

Dwarf surf clams (*Milinia lateralis***).**
Less than 1 inch long.
Predators: sanderlings, gulls, crabs and whelks.

Eastern oyster (*Crassostrea virginica***).**
6 inches long.
Grows in beds in calm waters, usually in salt marshes.
Predators: birds, oyster drills and humans.

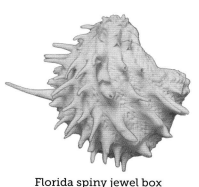

Florida spiny jewel box (*Arcinella cornuta***).**
2.5 inches long.
Why are jewel boxes so spiny?*

*Answer: to protect them from gastropod predators

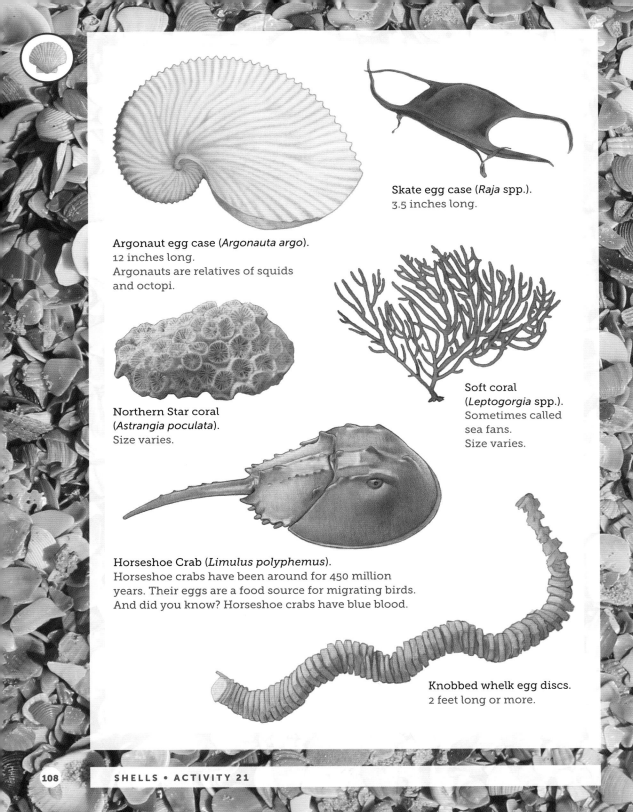

Argonaut egg case (*Argonauta argo*).
12 inches long.
Argonauts are relatives of squids
and octopi.

Skate egg case (*Raja* spp.).
3.5 inches long.

Northern Star coral
(*Astrangia poculata*).
Size varies.

Soft coral
(*Leptogorgia* spp.).
Sometimes called
sea fans.
Size varies.

Horseshoe Crab (*Limulus polyphemus*).
Horseshoe crabs have been around for 450 million
years. Their eggs are a food source for migrating birds.
And did you know? Horseshoe crabs have blue blood.

Knobbed whelk egg discs.
2 feet long or more.

Why Are Shells So Colorful?

Scientists know which chemicals cause certain colors in shells. For example, chemicals known as carotenoids* give shells a red or yellow color. Melanins give shells a brown color, and so on. What isn't well understood is the purpose of color in shells.

One theory is that color is used for camouflage, to help the animal blend in with a background and hide from predators. Another theory is that color is used for communication. Still another explanation is that color in a shell gives the calcium carbonate greater strength for resisting waves or the attacks of a predator.

Some scientists believe the color in some shells has no purpose, that it is only a by-product of what a mollusk eats or perhaps the result of waste products from digested food that get placed in the shell.

It seems likely that different mollusks **evolved** to use colors in different ways. Perhaps a young mind, intrigued by the beautiful array of colors in shells, is at this very moment pondering the issue and will one day do the research to explain why shells are so colorful.

Keyhole sand dollars (*Mellita quinquiesperforata*) are echinoderms, related to sea urchins. They graze the sea floor to feed on algae, worms and organic debris. In turn, they are fed on by fish, birds, octopi and sea stars. 4 inches in diameter.

*Carotene is a kind of carotenoid that gives carrots their orange color.

Black Shells

Sea Shell, Sea Shell,
Sing me a song, O Please!
A song of ships, and sailor men,
And parrots, and tropical trees . . .
—Amy Lowell (American poet)

Black shells (or black-stained shells) are found on nearly every beach in the Carolinas. In fact, they are found on beaches all over the world. In almost every case the black staining of the shell occurred after the organism's death. It is rare to find living shells stained black.

This is a two-part activity. The first part is to determine the percentage of black shells on a beach and to establish what types of shells become black. The second part is an experiment to learn how shells blacken.

Counting Black Shells

WHAT YOU NEED Energy, a notebook and a seashell field guide.

WHAT TO DO Draw a 3 × 3-foot square on the beach. Count the number of blackened and nonblackened shells in your square.

Black-stained shells.

Include in your count both large shell fragments and whole shells. Record your numbers in a notebook. Note the names of the shells as well. Repeat this activity at different locations.

On a natural beach, 5–15 percent of shells are typically black. However, some beaches, especially replenished ones (see Activity 31), have a much higher percentage of blackened shells. About what percentage of shells at your beach are black (see next activity for help with percentages)? Which species of shells are most often black? Are there some shells that never turn black?

Make Your Own Black Shells

WHAT YOU NEED A shovel and shells with their original colors.

WHAT TO DO Gather some fresh seashells (or shell fragments) and bury them in mud for several weeks. The shells can be in a bucket or buried a foot deep in a salt marsh or in muddy soil. The point is to prevent the shells from being exposed to the air. After 4–6 weeks or longer dig the shells up and see if any color changes have occurred.

Can you see a difference in the rate of blackening in different species? Where in the natural coastal environment would you find a muddy environment to blacken shells?

Where and How

Black staining occurs when iron combines with sulfur in tiny spaces within the shells to form microscopic crystals of black pyrite. This happens when no oxygen is present, as for example when the shell is buried in mud. You may have observed that shells blacken at different rates. The jingle shell blackens within a few days (we think). Most clams and snails blacken within six weeks, and scallops apparently don't blacken at all.

Because there is no mud on beaches, shell blackening must originate somewhere else. On barrier islands black shells come from old salt marsh deposits that were once behind the island but ended up on the beach as the barrier island moved over the marsh (see Activity 11). Replenished beaches often have abundant black shells, which come from marsh deposits offshore, left behind when sea level rose over them and the shoreline moved inland. Black shells on barrier islands are proof of island migration.

Brown Shells

She sells seashells by the seashore.
If she sells seashells by the seashore,
She must sell seashore shells.

—Terry Sullivan (in honor of Mary Anning, a pioneer paleontologist)

A natural Carolina sand beach is usually a brown to yellow-brown color if you look down the length of the beach. This is caused by the presence of so many brown shells and by brown sand. Look closely at the shells and you will find an irregular brown staining (like rust stains on steel) on many shells, easily distinguishable from the natural shell colors.

Brown-stained shells and black-stained shells (Activity 22) are common on most beaches in the Carolinas, and in both cases the staining occurred after the animal's death. The difference between the two shell stains is that black shells changed color while buried and not in direct contact with the air. Brown shells, on the other hand, got their color while exposed on the beach in full contact with the air. A combination of iron and sulfur (iron sulfide) forms the black stain, and a combination of iron and oxygen (iron oxide) forms the brown stain.

Brown staining occurs when shells are in contact with air.

The purpose of this activity is to see how abundant brown shells are on a beach, to determine which shells become brown and which don't and to understand the significance of brown shells.

Counting Brown Shells

It might be a good idea to do the black shell counting activity (Activity 22) and this one simultaneously since the activities are similar.

WHAT YOU NEED Energy, a notebook and a seashell field guide.

WHAT TO DO Draw a 3 × 3-foot square on the beach where shells are common. Count the number of brown-stained shells (including large shell fragments) and non-brown-stained shells. Record your numbers in a notebook. Note the names of the shells as well. Repeat this activity at different locations. What percentage of the shells at your beach are brown (that is, for every 100 shells in your squares, how many are brown stained)? For example, if you counted 100 shells and 7 were brown, then:

$$\frac{7 \text{ brown-stained shells}}{100 \text{ total shells}} = 7 \text{ percent brown-stained shells}$$

Are there some kinds of shells that never turn brown? What can brown-stained shells tell us about the history of a shoreline?

Brown Shells and Former Shorelines

Most shells can turn brown, but a few, such as the common oysters, are rarely brown stained. On a natural beach most shells are brown, but on a replenished beach, black shells often predominate. Brown staining takes a long time, perhaps a matter of years, so an exercise where you stain your own shells brown would be difficult (but not impossible if you have the patience).

Occasionally brown shells are found on the continental shelf off the Carolinas, some as deep as 300 feet or so. These shells represent former shorelines that existed during times of lowered sea level (see Activity 9).

Examples of shells that have been partly or completely stained brown.

Shell Roundness

The sea—this truth must be confessed—has no generosity.
—from Joseph Conrad, *The Mirror of the Sea*

A seashell is soft. You can easily scratch it with a pocketknife. But quartz, the main component of beach sand, is much harder. You can scratch a knife blade with quartz. As shells are tumbled in the surf zone they rub against sand grains and against each other and become rounded (a process known as mechanical weathering). Bigger waves mean more weathering and therefore rounder shells. Just remember, the bigger the pounding the more the rounding.

Roundness (measured on shell fragments, not whole shells) is defined by how smooth the edge of a shell fragment is. A roundness scale can be used to calculate smoothness. To do this, hold a shell fragment in your hand and compare it with the roundness scale. Then give the fragment a number. A higher number corresponds to a smoother fragment. Category 1 is what a shell might look like if you smashed it with a hammer. Category 4 shells feel almost perfectly smooth.

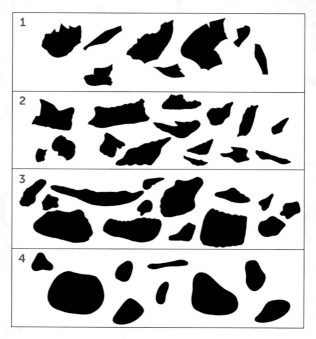

Classifying Shell Roundness

WHAT YOU NEED The roundness scale, a notebook to record your observations and a sense of curiosity.

WHAT TO DO Wander around the beach picking up shell fragments and classify each fragment using the roundness scale. Do this for 50 fragments. Which category do most of your shells fall in? If possible walk down the beach and repeat the exercise. If there is a beach on the sound side of the island you are visiting, you might try the activity there as well.

BRAINTEASERS

Cape Hatteras, N.C., has large waves and frequent storms, while Hilton Head or Hunting Island State Park in South Carolina have smaller waves and less frequent storms. How would you expect the roundness of shell fragments to differ on the two islands?

Many beaches have been replenished (Activity 31) with sand and shells pumped on the beach from local inlets or from the continental shelf. What would you expect the shell roundness on replenished beaches to be like?

Inevitably, there are shells on most beaches that look like they were recently smashed and broken. What do you think breaks up these shells?

More about Shell Roundness

Shells on a beach with typically high waves (like Cape Hatteras) generally have more rounded shell fragments than a South Carolina beach with lower wave energy. But sometimes humans interfere with the natural processes of weathering. Shells on a replenished beach are often sharper (less rounded) than those on a natural beach because the dredges pumping sand onto the beach tend to break the shells into jagged pieces.

Crashing waves, especially during storms, can break up shells. Another common cause of broken seashells are four-wheel-drive vehicles. Nature breaks up shells too. Stingrays, birds like oystercatchers and certain crabs can smash shells to obtain dinner. Other critters drill through or weaken the shells in other ways that hasten the weathering process. (See Activity 30.)

Did you know? People make necklaces, bracelets or mobiles out of well-rounded shell fragments. Maybe you can make some beach jewelry for your friends and family.

Shell Orientation

To me the sea is a continual miracle.

—Walt Whitman (American poet)

Bivalves like clams, scallops or oysters are positioned on the beach in one of two ways. The space where the organism once lived either faces up (concave up) or it faces down (concave down). One of these orientations is much more common than the other. This activity demonstrates which orientation is usually seen on a beach and explains why. You can do the activity with most bivalves, but it works best with large clams such as the southern quahog (*Mercenaria mercenaria*) or the Atlantic giant cockle (*Dinocardium robustum*) (see Activity 21).

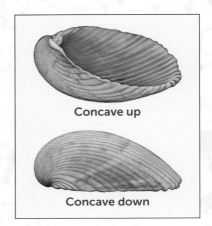

Concave up

Concave down

Which Way Is Up?

WHAT YOU NEED A notebook, a pencil and a beach with shells.

Concave up	Concave down
ՊՊ ||||	//

WHAT TO DO Walk along a beach where there are plenty of shells (low tide might be a good time for this activity). Look for clams or scallops. Every time you encounter a shell note whether it is concave up or concave down and record the count in a notebook. Count at least 25 shells. Which orientation is the most common?

Shell Orientation in the Water

Now that you have determined which way most shells are positioned on the beach, do this experiment to understand why.

WHAT YOU NEED A clam shell, a beach with small surf, clear water that's a foot or more deep and a face mask (optional).

WHAT TO DO Drop a clam shell into clear, calm water. This can be accomplished by wading into the surf or by using a 5-gallon bucket of water. When your shell settles on the bottom, note its orientation. Is it concave up or down? Change the initial position of the shell as you drop it to see if that changes the outcome.

   Next, take a clam shell and wade ankle-deep into the surf. When the waves recede, place the shell concave up on the sand. When the next wave comes in, what happens? Is the orientation of your shell the same or different?

   Another approach is to drop a shell into the surf and use a face mask to observe what happens. What is the orientation of the shell when it settles on the sand? Keep watching with your face mask to see if the shell changes its orientation. It likely will do so if the energy from the breaking waves is strong enough. *Note*: Don't do this activity in the surf during a storm or if the waves are too high.

Explaining Shell Orientation

The stable orientation for shells falling through the water is concave up, yet shells on a beach are mostly concave down. That's because waves easily turn over shells that are pointing up but have a hard time flipping shells that are pointing down. Therefore, the most stable position for shells on the beach and in the surf zone is concave down. Now, if you could swim 20 miles offshore and dive to the sea floor, you would find many of the clam shells oriented up. That's because there is less wave energy on the sea floor in deeper water than in the surf zone.

Geologists can use the orientation of fossil shells in rocks to determine whether the sediments were deposited near the beach or in deeper water. Deep-water sediments from the continental shelf have a mix of shells pointing up as well as down. Most shells in beach sediments point down.

EXPLORE YOUR WORLD

There is another kind of shell orientation you might see at the beach—the position of long, narrow shells (oysters, razor clams, etc.) relative to the surf. Can you find a preferred orientation for these shells? Do an experiment like the one in this activity to explain why such an orientation exists. Perhaps there are orientations of shells, driftwood, fossil sharks' teeth and the like still waiting to be discovered.

Shell falling through the water.

Shells on sea floor in the surf zone.

Fossils

I think it inevitably follows, that as new species in the
course of time are formed through natural selection, others will
become rarer and rarer, and finally extinct.
—Charles Darwin, *On the Origin of Species*

A kid was on his hands and knees, head bent close to the sand in intense
but joyful concentration. Sometimes he'd pick something up from the
beach, examine it for a moment, and either toss it aside or drop it into a
plastic bag.

"What're you looking for?" asked another boy walking up from the
water.

"Fossil sharks' teeth," said the kid. "They're everywhere here."

"What do they look like?"

"Well, they're usually black and curved, sometimes with jagged edges."

The boy joined the search, whooping with delight upon finding his first
shark tooth.

"That's a big one," said the kid. "Not bad for a first find."

A girl approached. "What're you looking for?

"Fossil sharks' teeth," answered the kid. "They're pretty cool."

"Is this one?" she asked, holding up something for him to examine.

He paused, ransacking the pockets of memory. "It's a mako tooth,
I think."

"Awesome," the girl said. And then she joined the search.

Others came. Before long a dozen kids and a sprinkling of adults crowded
the high-tide line. All afternoon they combed the sand for sharks' teeth
until the light grew dim and the westering sun became a fading ember
glowing through the trees.

Finding Sharks' Teeth (and Other Fossils)

As the story implies, looking for fossil sharks' teeth is fun (and addictive!). We have found sharks' teeth on nearly every beach in the Carolinas. But there are other kinds of fossils that wash up on the beaches as well. There are ancient shells, mostly between 10 and 80 million years old, which eroded out of rock outcrops on the continental shelf and later were washed up on beaches. Other fossil shells are more recent and are hard to tell apart from modern shells on beaches. These probably range in age from a few hundred to a few thousand years old and can appear to be in very fresh condition. Oyster shells on beaches are an example of this kind of fossil. Much rarer beach fossils are the remains of Ice Age land animals and marine animals (of various ages). These include huge megalodon teeth, whale and dolphin vertebrae, crabs, turtles and more exotic (but much rarer) land fossils such as mastodons (extinct elephants), bison, saber-toothed cats, horse, deer and giant armadillos.

WHAT YOU NEED A small plastic bag and a fossil field guide.

WHAT TO DO You can find sharks' teeth anywhere on the beach, but we recommend searching the intertidal zone, especially at low tide. Look for small black or brown, non-shelly objects from ¼ to 1 inch or more in size. In recent years hunting for fossil sharks' teeth has become increasingly popular from Virginia to Florida. Three good locations in the Carolinas for finding fossils are mentioned on the next page.

How many different kinds of shark teeth fossils can you find in the Carolinas? Why are fossil sharks' teeth black? How big was a megalodon? Are megalodon still swimming in the sea today?

Edisto Beach, South Carolina. One of the best beaches for fossil hunting in the Carolinas. The fossils are from three different time periods: Pliocene and Miocene marine fossils and Pleistocene land mammals (see the geologic time scale).

Myrtle Beach, South Carolina. A productive fossil hunting site (including North Myrtle Beach), probably because so many beach replenishment projects have brought fossils to the beach from the continental shelf.

Topsail Island, North Carolina. Famed for sharks' teeth and giant fossil oysters.

Great white shark (*Carcharodon* spp.). 1–1.5 inches long.

Snaggletooth shark (*Hemipristis serra*). 0.75 inches long.

Tiger shark (*Galeocerdo* spp.). 0.75 inches long.

Dolphin tooth. 1 inch long.

Cow shark (7 gill) (*Notorynchus* spp.). 1.5 inches long. Rare on beaches.

Megalodon (*Charcharodon megalodon*). 3–7 inches long.

Sand tiger shark (*Charcharias* spp.). 1–1.5 inches long.

Hammerhead shark (*Sphyrna* spp.). 0.5 inches long. Look for the groove at the top of the crown.

Mastodon tooth.
7 inches long.

Giant fossil oysters
(*Crassostrea gigantissima*).
10 inches long.
Topsail Island, N.C.

Land mammal bones.
4 inches long.
Edisto Beach, S.C.

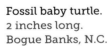

Crocodile tooth.
1–3 inches long.

Fossil baby turtle.
2 inches long.
Bogue Banks, N.C.

Coquina (limestone made
of fossil shells).

More about Beach Fossils

A good introduction to the world of fossil collecting would be to visit the Aurora Fossil Museum. Spend a few hours collecting fossils from their outdoor digging pit. Then try your luck at the beach.

There are hundreds of fossil shark species, most of which are extinct (including the megalodon). Sharks are constantly losing their teeth and then regrowing new teeth. If a tooth gets buried in sediment, it will likely become a fossil. After thousands of years, the surrounding sediment changes the chemistry of the tooth. Its original white color becomes black, gray or brown.

How big was a megalodon? Judging from its teeth, which is all that has survived in the fossil record, it may have reached 50 or 60 feet long—big enough to swallow today's great white and still be looking for dessert.

6

Life at the Beach

On the beach you can live in bliss.
—Dennis Wilson of the Beach Boys

A beach is a hard place for life to take hold. Waves, winds and tides are constantly shifting sand from one place to another. Gentle summer waves tend to push sand landward, creating a wide, flat beach, while heavier surf in winter removes that sand offshore, making the beach narrow and steep. Storms can reconfigure an entire beach in a matter of hours. The sand itself is a problem; it doesn't retain rainwater and under the hot summer sun the sand can reach unbearable temperatures. The few plants that manage to survive the dry, salty, desertlike conditions on the upper beach are too low to the ground to provide much shelter for the animals living there.

Yet beaches are full of life. Birds, turtles, grasses, trees, lizards, crabs, clams, snails, insects, fish and many other life forms live or nest on the beach. The activities in this chapter illustrate how these organisms have adapted to the harsh, unstable environment of a Carolina beach. Special attention is given to the meiofauna, critters too small to see with the naked eye that live between sand grains in the surf zone. These critters are the base of a long food chain that eventually leads to humans.

Life in the Fast Lane

Walking back against the flats of that Georgia beach,
I was always aware I was treading on the thin rooftops of an underground city.
—Rachel Carson (American writer and environmentalist)

Every time you step into the surf you are treading on thousands of creatures living in the sand. Some are big enough to be picked up by your hand—clams, worms, sand dollars. Others are just visible to the human eye. Most are so small you need a microscope to see them.

Digging for Mole Crabs

Mole crabs (*Emerita talpoida*) are harmless to people. Unlike other crabs, they have no claws and always move backward (never sideways or forward). They tend to cluster in groups, especially in late spring and early summer. You can detect their presence by watching the swash retreat back to sea. When mole crabs extend their antennae to collect food, they form a V-shape in the receding water. Mole crabs migrate up and down the beach with the tides. When the tide goes down, they "surf" on receding waves to get to deeper water. Aside from being an object of curiosity and fun for kids, mole crabs are an important food source for a variety of critters (mostly birds and fish).

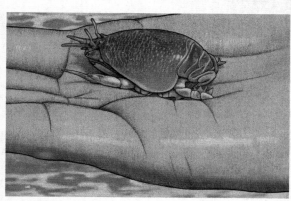

Mole crab

WHAT YOU NEED Bare hands and a stout heart.

WHAT TO DO Dig a hole with your hands about eight inches deep in wet sand at the edge of the swash. Let seawater fill in the hole; swirl the water around with your hands and look for mole crabs burrowing deeper into the sand. Pick up a mole crab and look for the eye stalks and antennae. How many legs does a mole crab have? Are there any pink masses of eggs on its belly? When you are finished with your examination, gently place the mole crab on the sand. Watch how it furiously burrows to safety by digging backward. What do you think happens to mole crabs and other life living amongst the sand grains when a beach is **replenished** (Activity 31)?

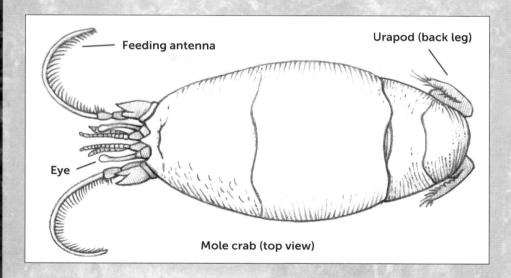

Feeding antenna

Urapod (back leg)

Eye

Mole crab (top view)

FUN IN THE SUN

Coquina clams (Activity 21) are small, colorful clams often found in the sand next to mole crabs. Try picking up a live coquina; put some sand in your hand and watch as it tries to wiggle its way through the sand.

Meiofauna

The word "meiofauna" (*my*-oh-fawna) means "the lesser animals."
The largest are just visible to the human eye, but most can only be
seen through a microscope. Meiofauna are found in sediments in all
of the world's watery habitats—rivers, lakes, marshes and even deep
ocean trenches. A square yard of beach sand contains around a million
meiofauna, almost all in the upper inch or two of sand.

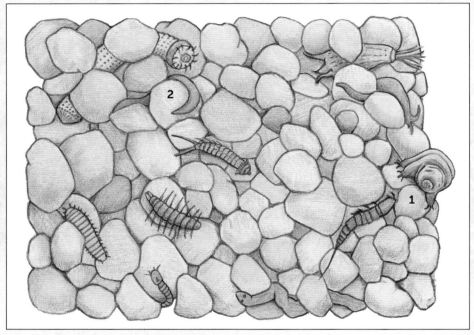

Meiofauna live between grains of sand at the beach. The handsome
critter on the far right (1) is a water bear. Round worms (2) are the most
common kind of meiofauna.

Lesser but Great

The miniature kingdom that exists between sand grains at a beach is amazingly diverse. Thousands of species and at least 22 major groups of animals (phyla) live there. By comparison, tropical rainforests support only about 16 major groups. But the surf zone is a harsh environment. The relentless pounding of waves and flushing of tides means sand grains are constantly rubbing against each other, and the space between the grains alternates between periods of wet and periods of dry. Some meiofauna have evolved armor to protect their bodies from abrasion. Most have become long and flattened to better slide between the grains. Others have developed claws, suction cups or glue to hold fast to grains while seawater rushes through the sand. Water bears can remain dormant and dehydrated when out of water, waiting for years until bouncing back to life again when wetted.

Like mole crabs, meiofauna occupy the base of a food web that supports larger, more familiar animals like fish, crabs, shrimp, mollusks, birds, sea mammals and humans. But meiofauna provide another important service to the beach community. In addition to preying on each other, they feed on detritus (organic remains), plankton, bacteria and algae. Thus they keep the beach clean and rid the sand of potentially harmful microbes. They are, as it were, the sanitary engineers of our shorelines.

Meiofauna are sensitive to pollution, and scientists monitor their numbers to determine ocean pollution levels. Because beach replenishment requires dumping tons of sand on a beach, the meiofauna and other critters living there are killed, turning the beach into a sterile wasteland. The birds and crabs must hunt elsewhere for their dinners.

Birds at the Beach

In order to see birds it is necessary to become part of the silence.
—Robert Lynd (Irish writer)

Try the following experiment. Find a secluded beach removed from the rattle and rush of urban life. Sit by the edge of the dunes, close your eyes and listen. What do you hear? Depending on local conditions, you will likely hear the sound of waves, the whistle of wind across sand and the cry of gulls. For most of us, waves, sand and the sounds and sights of gulls are what come to mind when we think about beaches. But gulls and the hundreds of other seabirds that frequent the shore are more than just a backdrop for a human-centered beach experience. They play a crucial role in coastal ecology by consuming vast quantities of fish, crabs, shrimp and clams. In turn, their eggs provide food for a variety of other critters, including foxes, raccoons and crabs.

The focus of this activity is on those birds common to the area between the dunes and the immediate offshore waters. We are lucky that both South and North Carolina offer some of the best places to observe seabirds on the East Coast. Unfortunately, because of habitat loss and other factors, many species are in sharp decline, their numbers decreasing year by year.

How many birds at the beach can you identify? What do they eat? Where are the best beaches to see birds? Are there some birds that spend their entire lives over water, or do birds have to return to land to rest from flying?

How Birds at the Beach Get Their Dinner

Each seabird has a unique feeding strategy (some may use more than one).

Here are some strategies you might observe at the beach:

1 *Surface diving*. Birds like loons or cormorants dive while floating on the water's surface.
2 *High diving*. Some birds dive while in flight and either catch fish near the surface or chase their prey deep underwater.
3 *Feeding in flight*. While in flight some species grab fish with their beaks or catch fish in their talons.
4 *Piracy*. Some birds steal food from other birds.
5 *Waders*. Sanderlings, herons and other species probe the swash for mollusks, fish or mole crabs.
6 *Scavenging*. Some birds eat animal remains, plants and other organic debris washed ashore in the wrack.
7 *Predation*. Birds of prey like hawks and falcons eat small birds, mammals, etc. Usually part-time residents or visitors.

WHAT YOU NEED   A notebook and pencil, binoculars, a field guide and the patience to observe birds over hours and days.

WHAT TO DO   Identify as many birds as you can. Observe their feeding habits. Make a chart to classify birds according to how and where they feed. Why do most birds only feed at one particular place? Why don't they feed anywhere that food is available?

Date	Bird	Feeding strategy	Location
6/1/14	Laughing gull	Scavenger	Wrack line, Folly Beach, S.C.
6/1/14	Brown pelican	?	?

Make a chart like this one to record how and where birds feed at the beach.

Ecological Niche

Imagine what would happen if people in a city could eat at only one restaurant. The lines would be impossibly long. Everyone would be hungry and irritable, pushing and shoving in line. Fortunately, cities have lots of restaurants. If one restaurant gets too crowded, people eat somewhere else. It's the same with nature. An environment like a beach has multiple sources of food. Animals can feed at many different locations. This helps reduce competition between species (and keeps the birds from shoving and pushing in line). The role of an organism, including where it lives and what it eats, is called its **ecological niche**. Each plant or animal occupies a particular niche. An example of a niche is that of a predator feeding on fish in shallow coastal waters. The great egret fits this niche well because it **evolved** a sharp, bladelike bill useful for spearing fish.

Great egret (*Ardea alba*).
40 inches tall.

Red-breasted merganser (*Mergus serrator*). 20–26 inches long. Top: male; bottom, female. The male and female of many (though not all) bird species can be very different in appearance. Red-breasted mergansers occur by the thousands on Pamlico Sound, N.C.

Popular Birding Sites on the Coast
North Carolina
Pea Island National Wildlife Refuge
Cape Hatteras National Seashore
Cape Lookout National Seashore, including:
Core Banks
Ocracoke Island
Shackleford Banks
Fort Fisher
Sunset Beach
South Carolina
Cape Romain National Wildlife Refuge
Huntington Beach State Park
Hunting Island
Edisto Beach State Park
Pinckney Island National Wildlife Refuge

DID YOU KNOW? ───

Huntington Beach State Park is considered by some to be one of the best birding spots on the East Coast.

Bonaparte's gull
(*Larus philadelphia*).
12 inches long.

Ring-billed gull
(*Larus delawarensis*).
19 inches long.

Great black-backed gull (*Larus marinus*).
28–31 inches long. Wingspan: 66 inches.
Largest gull in the world. Lesser black-
backed gulls (*Larus fuscus*) are smaller
than great black-backed gulls.

Laughing gull
(*Larus atricilla*).
16 inches long.

Herring gull (*Larus argentatus*).
23–26 inches long. Yellow bill
with red spot on lower tip.

Double-crested cormorant
(*Phalacrocorax auritus*).
33 inches long. Wingspan: 4 feet.
Often seen spreading wings to dry.

Brown pelican
(*Pelecanus occidentalis*).
41 inches long.
Wingspan: 7 feet. White
pelicans (*Pelecanus
erythrorhynchos*) are
occasional visitors to
Carolina beaches.

Osprey
(*Pandion haliaetus*).
24 inches long.
Wingspan: 6 feet.
Often seen near
inlets, where they
nest on channel
markers.

Oystercatcher
(*Haematopus palliatus*).
17 inches long. Nests on
undisturbed beaches.

Ruddy turnstone (*Arenaria interpres*).
9 inches long. Like some other birds, the ruddy turnstone changes color when breeding. Right: breeding plumage. Left: Nonbreeding plumage.

Wilson's plover (*Charadrius wilsonia*).
7–8 inches long.
Nests on beaches and dredge-spoil islands.

Marbled godwit (*Limosa fedoa*).
16–20 inches long.

Sanderlings (*Calidris* spp.).
6–9 inches long. Sanderlings breed in the Arctic and spend summers in South America.

Willet (*Tringa semipalmata*).
14–17 inches long.

Black skimmer
(*Rynchops niger*).
16–20 inches long.

Common loon (*Gavia immer*).
28–36 inches long.

Least tern (*Sternula antillarum*).
9 inches long. Terns are small-
to medium-size birds with long
migration routes. The Arctic tern
(*Sterna paradisaea*) migrates from
Arctic to Antarctic waters and
back, a distance of 40,000 miles.
Least terns sometimes nest on
gravel rooftops where their normal
beach nesting grounds have been
eliminated by development.

Fish crow (*Corvus ossifragus*).
16–20 inches long.
Slightly smaller than the American
crow and has a different call.

White ibis (*Eudocimus albus*).
22–27 inches long.

Snowy egret
(*Egreta thula*).
20-27 inches long.
Nests in trees near
water.

Reddish egret
(*Egreta rufescens*).
30 inches long.

Wood stork
(*Mycteria
americana*).
34–47
inches long.
Wingspan:
more than 5
feet.

Great blue heron (*Ardea herodias*).
Largest North American heron.
45–52 inches long. Wingspan: 70 inches.

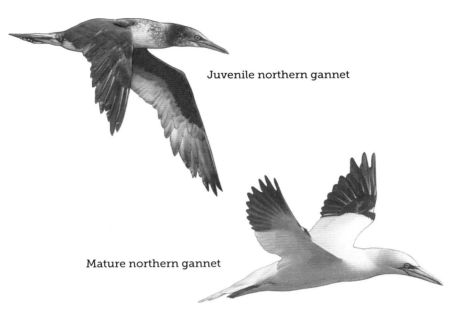

Juvenile northern gannet

Mature northern gannet

Northern gannet (*Morus bassanus*).
38–43 inches long. Large white bird with a wingspan of nearly 7 feet. Like most birds, the juvenile or immature gannet does not resemble the adult. Gannets are winter residents in the Carolinas, often seen in the waters off the Outer Banks and in Charleston Harbor.

Red-winged blackbird (*Agelaius phoenicius*).
7–9 inches long. Not all birds at the beach feed on seafood. Some, like the red-winged blackbird and grackles, feed on insects or the seeds of sea oats.

Gulf Stream Pelagic Birding

Pelagic (puh-*lah*-jick) birds are those birds that spend most of their lives on the open ocean, rarely coming to land except to breed. The convergence of the north-flowing Gulf Stream and the south-flowing Labrador current 20–40 miles off the coast of Cape Hatteras creates rich fishing grounds, attracting commercial fisherman as well as seabirds. Local fishing boats organize frequent summer and winter cruises for birders wishing to spot birds impossible to see from land. Examples include shearwaters, puffins, South Polar skuas and storm petrels. Inquire at Oden's Dock in Hatteras Village or at Pirate's Cove Marina in Manteo or go to www.seabirding.com.

Feeding gulls (or any wildlife) is a bad idea. It encourages gulls to gather in large numbers, allowing disease to spread from bird to bird. Also, people food lacks the nutrients the birds need. Better give your leftovers to your local ghost crab. . . . Just kidding!

ACTIVITY 29

Plants and Salt

Green is the prime color of the world,
and that from which its loveliness arises.
—Pedro Calderon de la Barca
(seventeenth-century Spanish poet)

You would think beaches in the Carolinas would be ideal environments for plant growth. After all, they provide ample sunshine and plenty of rain. Yet the stretch of sand between dune and tide is nearly empty of plant life. Why is this so?

Certainly, hurricanes and other storms play a role. Strong winds and high seas can topple trees and uproot smaller plants. But plants are absent even on the beaches like Hilton Head, where hurricanes are rare. Something other than storm frequency must therefore hinder plant growth.

Two things stop plants from colonizing beaches: sand and salt. Sand lacks the nutrients for healthy plant growth and doesn't retain enough rainwater, in effect turning a beach into a coastal desert. Some salts are poisons that interfere with the chemical processes by which a plant makes food and grows. Salts also prevent plants from taking in enough water. You can tell when a plant is salt damaged by the "burned" look of its leaves.

Salt at the beach comes from the sea, either as wind-borne spray or by waves, tides and storm surges. The few plants that have evolved to live in salty environments are known as halophytes (*hal*-oh-fights) or "salt lovers." While driving to the beach you probably passed a common halophyte called spartina (*spar*-tine-nuh), happily growing in a salt marsh.

This activity is about how salt determines the shape, size and location of plants at the beach and why "salt loving" plants are important to the coastal environment.

Salt Tolerance

WHAT YOU NEED A beach with at least some **maritime forest**, dunes and natural beach vegetation; a plant field guide; and a pencil and notebook. *Note*: This activity can't be done on rapidly eroding beaches like those on Hunting Island, South Carolina. When a beach erodes too quickly the normal sequence of salt-tolerant plants gets changed.

WHAT TO DO Walk from the forest to the beach. As you get closer to the dunes how do size, numbers and spacing of plants change? Draw a sketch in your notebook of the trees nearest the beach. What is different about their shape compared to the same trees on the mainland away from the coast? The plants that are most salt tolerant will be closest to the beach. Those that are less salt tolerant will be farther back in the forest. Make a list of at least five plants in order from least to most salt tolerant.

Your list may vary somewhat from beach to beach, especially where humans have disturbed the natural vegetation. The list will certainly vary with latitude. South Carolina, with its warmer, more subtropical climate, is home to species less common on the cooler beaches to the north. Look for **sea oats** and live oaks. Both are native to the Carolinas and grow along the entire southern coast from Virginia to Texas.

Least salt tolerant			→	Most salt tolerant
Edge of forest			Dunes	Upper beach
Live oak	?	?	Sea oats	?

How Salt Spray Affects Plants

As you walk from the forest to the beach plants become fewer in number, less densely packed and smaller. Trees become twisted, with wedge-shaped crowns and branches leaning crookedly away from the beach. That's because salt spray stunts the parts of a tree closest to the ocean, a process known as salt pruning.

Salt pruning is nature's bonsai. Trees close to the beach are sculpted into beautiful shapes by strong winds and salt spray.

Another Salt Lover

Marine algae (known to most of us as seaweed) are halophytes that are sometimes washed up on the beach by storms. Found from warm tropical beaches to the cold shores of Antarctica, marine algae vary in size from the microscopic to giant kelp as tall as a tree. Scientists split them into three main groups: red algae, brown algae and green algae. All three types are on southeastern beaches. See if you can find an example of each type on your beach.

BRAINTEASERS

Why do some plants do well on beaches and others don't? How do seaweed and sea oats help dunes survive? Why is marine algae crucial for sea life as well as for life on land?

How to Survive Salt, Sun and Strong Winds

Plants **evolved** various strategies for surviving the harsh beach environment:

- Some, like American sea-rocket, are succulents. Succulents have leaves that swell with rains and store rainwater for later, drier times.
- Crested saltbush thrives in salty landscapes by releasing excess salt back into the environment.
- Beach marsh-elder has a thick, waxy surface that resists salt pruning.
- Sea oats are champions of beach plant survivors. They have deep roots that anchor the plants against strong winds. They also have a working relationship (a **symbiosis**) with bacteria that live in their roots and supply the plants with nitrogen. This relationship is of great value in the nutrient-poor soils of a dune field.

Don't Wrack Your Brain

Seaweed (marine algae) grows offshore. Spartina grows in salt marshes. But often storm waves and tides bring algae and spartina to the beach where they become part of the **wrack** (debris washed up on a beach). Under the right conditions wrack can be a "seed" around which new dunes form. The same is true for young sea oat plants growing on the

upper beach (see Activity 8). Insects and crabs that feast on wrack are themselves eaten by birds and other animals. Wrack may not look like much, but it keeps the birds, crabs and dunes happy.

Wrack is part of the beach food chain and a "seed" for new dunes to form.

Sea oats (*Uniola paniculata*). Sea oats are protected from collecting because of their role in stabilizing dunes. Birds and insects eat the seeds.

Bitter panicgrass (*Panicum amarum*). Like sea oats, bitter panicgrass helps keep dunes in place and prevents erosion.

American beachgrass (*Ammophila breviligulata*). American beachgrass also keeps dunes from eroding. It is more common north of Cape Hatteras, while sea oats are more common south of the Cape.

Live oak (*Quercus virginiana*).
Found from Virginia to Texas. An
important member of the maritime
forest. Along with bald cypress and
cabbage palms, live oaks are some of
the oldest trees in the Southeast.

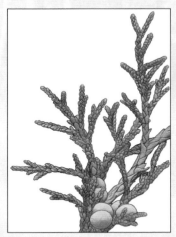

Eastern red cedar (*Juniperus virginiana*).
Found in eastern North America.
Has aromatic leaves and berries. Red
cedar near the dunes tends to get salt
pruned, looking more like a low bush
than a tree.

Longleaf pine (*Pinus palustris*).
Found from New Jersey to Texas.
Needles: 3 per bunch, 12–18 inches long.
Identified by its long needles, longleaf
pine used to cover much of North
Carolina. Only 10 percent of its original
range survives today. The oldest known
living longleaf pine tree is 460 years old.

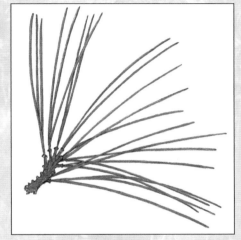

Slash pine (*Pinus elliottii*).
Found from southeastern South
Carolina to eastern Louisiana. Needles:
2–3 per bunch, 8–12 inches long.
Cones: 3–6 inches long.

Loblolly pine (*Pinus taeda*).
Found from New Jersey to Texas.
Needles: 3 per bunch, 5–9 inches long.
Cones: 3–5 inches long. Pine nuts
eaten by birds and squirrels.

Adam's needle (*Yucca filamentosa*).
Found in eastern United States. 2–3 feet tall.
Long stringy threads on leaf edges. Similar to
Y. gloriosa but not as tall.

Cabbage palm (*Sabal palmetto*). Found from Cape Fear, N.C., to Florida. 65 feet tall. Cabbage palms can live hundreds of years.

Prickly pear (*Opuntia* spp.).
Often growing in the dunes. The fruit of prickly pear is eaten by insects, birds, and people (after peeling off the thorns).

Spanish bayonet (*Yucca aloifolia*). Found from North Carolina to Gulf Coast. 6–12 feet tall. Yucca plants are pollinated by the yucca moth, which lays its eggs inside the plant's flowers.

Silver-leaf croton (*Croton punctatus*).
Found from North Carolina to Texas, where it's known as gulf croton. Leaves are grayish-green. Fruit resembles hard peas.

Russian thistle (*Salsola kali*).
Also known as tumbleweed. An invasive species (from Eurasia) made popular by countless Hollywood Westerns.

Beach-marsh elder (*Iva imbricata*).
Grows on dunes and upper beach.

American sea-rocket (*Cakile edentula*). Found from South Carolina to New England. Thick leaves with many small lobes. Small white flowers.

Seabeach amaranth (*Amaranthus pumilus*). A threatened species protected by the federal government.

Beach pennywort (*Hydrocotyle bonariensis*). Beach pennywort is a pioneer plant that boldly grows where few plants have grown before. Found from behind the dunes down to the upper beach.

Indian blanket (*Gaillardia pulchella*). A common sight along roads in the American Southwest. Also native to the Carolina coast.

Beach morning glory
(*Ipomoea pes-caprae*).

Beach evening primrose
(*Oenothera drummondii*).
Most primroses open their flowers at
night and are pollinated by moths.

Beach
vitex (*Vitex
rotundifolia*).
Invasive
species from
Asia that is
pushing aside
native plants.

Groundsel tree
(*Baccharis halimifolia*).
Yellow male flowers, white female
flowers. Grows in the coastal plain
and Piedmont of the southeastern
United States.

Sea purslane (*Sesuvium portulacastrum*).
A succulent found on beaches
worldwide. Native Americans used to
eat it after cooking it in several changes
of water, to reduce the plant's bitterness.

Beach plum (*Prunus maritima*).
According to the U.S. Department
of Agriculture, beach plums range
from Virginia to Maine, but we have
found the plant on beaches south of
Charleston.

Yaupon holly (*Ilex vomitoria*).
Fruit eaten by ducks, quail, wild turkeys,
raccoons, bears, deer and foxes.
Grows in the dunes.

Virginia creeper
(*Parthenocissus quinquefolia*).
Usually seen as vines growing
up tree trunks in the forest, but
it also "creeps" onto the upper
beach. Berries are poisonous to
people but are eaten by birds.

Crested saltbush (*Atriplex cristata*).
Eaten by butterflies, moths and people.
Found throughout the dunes.

ACTIVITY 30

Murder Mystery

When you have eliminated the impossible,
whatever remains, however improbable, must be the truth.
—Arthur Conan Doyle, *Sherlock Holmes*

A crime has been committed. There's been
a murder at the beach. A serial killer is on
the loose. Another mollusk has been found
with a hole savagely drilled through its shell,
its innards sucked out, its empty carcass
wantonly abandoned in the surf. Your task is
to identify the killer.

Most of us see the beach as a place of gentle beauty, of pretty sunsets,
colorful shells and quiet meditation of surf and wind. We forget that
underlying all that natural beauty is a vibrant ecosystem where life
engages life in a fierce and brutal competition for survival. Plants compete
for nutrients and sunlight. Animals compete for the limited amount of
food provided by plants. Predators compete for limited numbers of prey.

Life feeds on life, a principle well illustrated at the beach, especially on
the surface of seashells. Examine a hundred shells and you will likely find
some with holes, scratches, chipping, nicks or grooves—the marks of a
hungry predator, intent on devouring its hapless victim.

For this activity you will play the role of beach detective, a kind of
Sherlock Holmes by the sea. You will examine the clues left at the scene
of the crime (in this case on the shells themselves) to determine how the
organism was killed. Then with your superior intellect and advanced
powers of deductive reasoning you will identify the killer and solve *the
Mystery of the Murdered Mollusk.*

Shell Predators

Your list of suspects is a long one. Predators that feed on shellfish include other mollusks, sea stars, birds, fish, sharks, crabs, lobsters, worms and sponges and some mammals (and of course humans). Each predator has its own strategy for getting through the outer shell of its prey to the soft, juicy animal within. Some pry the shells open with brute force and others drill holes. Some practice chemical warfare using acids and poisons, and some mercilessly drop shells from heights, smashing them on the hard ground below.

WHAT YOU NEED A shell field guide (see Activity 21) and a notebook to record your observations. A magnifying glass may be useful but is not necessary.

WHAT TO DO Use your incomparable observation skills to examine shells along the edge of the surf for evidence of foul play. If you find a suspicious mark, like a hole or a chipped edge, compare the mark to the examples in this activity to determine if a murder has occurred and what the murder weapon might have been. That will assist in identifying the killer.

Not every broken shell is the result of an attacking predator. The large, jagged hole on the shell of this Atlantic giant cockle (*Dinocardium robustum*) was the result of being tumbled in the surf.

Large, jagged holes on whelk shells are the bite marks of loggerhead turtles.

Boring sponges are actually quite interesting. Some, like the yellow boring sponge (*Cliona celata*), secrete acids to create dozens of tiny holes in their prey.

Shark eye snails (*Neverita duplicata*) (also called moon snails) drill countersunk bore holes. The inner hole is noticeably smaller in diameter than the outer hole.

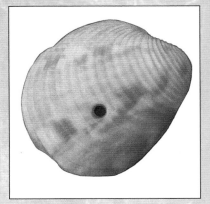

Atlantic oyster drills leave small, straight holes. Thick-lipped drills make small, slightly beveled holes.

Long grooves cut into the surface of a shell are made by bristle worms that use the grooves as living space. This weakens the shell, making it more easily damaged by surf.

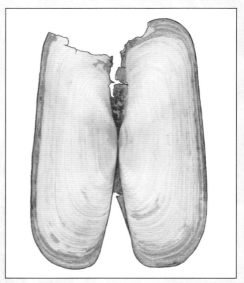

Chipped edges on clams are caused by whelks that use their own shells to pry open the clams.

This razor clam was probably damaged by a bird, the American oystercatcher (*Haematopus palliatus*).

Crows, herring gulls and ring-billed gulls drop shells from heights onto the beach and parking lots, smashing the shells and then eating the contents.

Crabs break open shells with their claws, sometimes leaving scratch marks on the surface.

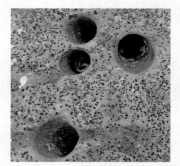

Bivalves like piddocks and *Gastrochaena* drill into rocks and thick shells to carve out their living quarters.

CHAPTER

7

The Environment

What does it profit a man,

if he shall gain the whole world, and lose his soul?

—Mark 8:36

There are two opposing views about our relationship to the earth. The first holds that nature exists for human benefit. To change, control or make money from nature is good because it improves people's lives. This has been the traditional view of engineers, developers and industry (fishing, mining, lumber, oil, etc.). The second view claims that plants and animals (and beaches) have a right to exist, regardless of their usefulness to humans. People must learn to live with nature and not tamper with natural systems. This is a philosophy shared by some of the world's religions, some Native American traditions, the teachings of St. Francis of Assisi and more recently, the movement known as Deep Ecology.

The first view is a human-centered way of seeing the world. The second is a nature-centered one. Most people find a position somewhere between those two extremes. In an age of global warming, rising seas and shoreline retreat we need to consider what our relationship to the shoreline is or should be. The question is an important one and is the focus of the activities in this chapter.

Replenished Beaches

In wildness is the preservation of the world.
—from Henry David Thoreau, "Walking"

Replenished beaches (also known as nourished beaches) are artificial beaches where sand lost to erosion is replaced by sand from another source. In the Carolinas, most sand for replenished beaches is taken from the **continental shelf** just offshore from the beach. The sand is sucked up from the ocean floor by a **dredge**, pumped ashore in a pipe and then smoothed with bulldozers. Sand is also pumped in from the **sound** on the back side of **barrier islands** or dredged from **inlets**. And occasionally it is brought in by the truckload from sites on the mainland.

Replenished beaches temporarily stop erosion and provide a wide beach for storm protection and recreation. But they are expensive and in the Carolinas have typical life spans ranging from two to four years. In addition, replenishment kills all the critters living within the beach. This activity explains how to know if a beach has been recently replenished or not and discusses the pros and cons of replenishment.

Beach replenishment involves pumping sand onto a beach.

How to Recognize a Replenished Beach

BEACH COLOR

A natural Carolina beach is a light-brown color. If the overall beach color is gray, it is certainly a replenished beach.

SHELL COLOR

Most shells on a natural beach will have their original colors, and brown-stained shells will be more common than black ones. A beach that has been replenished will likely have many black shells and possibly many white shells. If the replenished beach sand was delivered by a dredge, the shells will likely be broken and have sharp, jagged edges.

SHELL SPECIES

Replenished beaches vary a lot when it comes to the number of shell species on them. But if there are only two or three shell types present, the beach has almost certainly been replenished.

BEACH SCARPS

Scarps (cliffs of beach sand six inches to six feet in height) are a sign of rapid erosion. Since replenished beaches erode much faster than natural ones, scarps are more common on replenished beaches.

WHAT YOU NEED A beach and a keen eye.

WHAT TO DO Walk along the beach and look for signs of replenishment. Can you determine if the beach has been replenished or not? Sometimes it's hard to distinguish replenished beaches from natural ones. If all else fails, ask a lifeguard if the beach has been replenished!

How does replenishment affect animals living on a beach? What about the critters on the sea floor where dredging is done? We live in an age of rising seas and widespread coastal erosion. Some argue that replenishment is needed to keep beaches from washing away. Others say the beaches will eventually erode anyway and replenishment is a waste of time and money. What do you think?

Replenishment: Pros and Cons

THE GOOD

Replenishment creates a wider beach that can better protect houses from storms. A wider beach means a larger recreational beach with more space for people to enjoy. Replenishment is also good for the economy. Towns on the coast are supported by tourism. If the beaches erode away, tourists will go elsewhere. Motels, souvenir shops, surf shops, restaurants and other businesses will be ruined.

THE BAD

Offshore dredging kills life on the sea floor. Beach replenishment kills small animals (called meiofauna) living in the sand by burying them. Meiofauna are the base of a food chain that supports larger critters like crabs, birds and fishes. It may take three to four years for life to repopulate a replenished beach. Since Carolina beaches usually need replacement every three or four years, the beach ecosystem has no time to recover. This is why a replenished beach has fewer organisms than a natural one. Replenishment is also expensive, sometimes costing taxpayers millions of dollars for a single mile of beach. As sea level rises, the rate of coastal erosion is expected to increase, requiring even more money for replenishment projects.

THE UGLY

Ideally, the replenished beach sand should be similar to the original sand, but sometimes sand pumped up from the ocean floor is too muddy, shelly or rocky and makes an ugly beach. Shells on a replenished beach can be so sharp that walking with bare feet is hazardous.

Replenished Beaches in the Carolinas*	
South Carolina	**North Carolina**
Daufuskie	Ocean Isle Beach
Hilton Head	Holden Beach
Hunting Island	Oak Island
Edisto Beach	Caswell Beach
Seabrook Island	Bald Head Island
Kiawah Island	Kure Beach
Folly Beach	Carolina Beach
Sullivan's Island	Masonboro Island
Isle of Palms	Wrightsville Beach
Pawleys Island	Figure Eight Island
Debidue	Topsail Island
Garden City Beach	West Onslow Beach
Huntington Beach	Bogue Banks
Surfside Beach	Cape Hatteras
Arcadian Shores	Ocracoke Island (E)
Myrtle Beach	Hatteras Island (N)
North Myrtle Beach	Pea Island
Waites Island	Nags Head

*Developed by the Program for the Study of Developed
Shorelines at Western Carolina University.

The beaches listed above have been replenished multiple times, beginning with Wrightsville Beach in 1939. Probably the most expensive replenishment project was at the south end of Pea Island on the Outer Banks, costing 10 million dollars per mile in 2014.

EXPAND YOUR WORLD

Learn more from the internet or from books about the costs and benefits of replenishment. How do we balance the economic needs of society with the needs of organisms living on the beach? Start a discussion with friends and family.

Concrete Beaches

Man marks the earth with ruin,—his control stops with the shore . . .
—from Lord Byron, "Childe Harold's Pilgrimage"

Almost the entire shoreline of the Carolinas is eroding. Hundreds of buildings would have fallen in by now if engineers hadn't come to the rescue to "stabilize" the shoreline. Engineers use two different approaches to hold a shoreline in place. The "soft" approach is to pump sand on the beach (see Activity 31), which temporarily widens the beach. The "hard" approach is to stop the landward march of the shoreline using **seawalls** and **groins.**

Seawalls are walls built parallel to the shoreline on the upper beach (usually near the dune/beach boundary). They can be built of steel, concrete, rocks, wood, wire baskets filled with rocks or sandbags. In a few places in the world seawalls are made of junked cars or piles of wood and concrete from fallen buildings!

Groins are walls built perpendicular to the shoreline and are made of the same materials as seawalls. They trap sand moved by the **longshore current** (see Activity 4). The longshore current is a wave-generated current that usually flows from north to south in the Carolinas, though in fact it can flow in the opposite direction if the wind is southerly.

Seawalls are intended to block waves from damaging buildings or dunes and to hold the shoreline in place. But seawalls have a negative side. Over time they destroy the beach. This is why both North and South Carolina eventually made seawalls illegal. To learn about the negative side of groins do the following activity.

Groins

Longshore currents move sand up and down the shoreline. Anything that blocks the longshore current (like a groin) causes sand to pile up.

WHAT YOU NEED A beach with groins and an eye for details.

WHAT TO DO Walk up and down a beach where groins are present. Determine the dominant direction of the longshore current by noting on which side of a groin sand has accumulated. What happens to the beach on the **downdrift** side of a groin? If you lived on the beach, would you be happier with a house on the **updrift** side of a groin or the downdrift side? Do piers cause sand to pile up like a groin does? What is the purpose of a groin? Why are groins dangerous to swimmers at high tide?

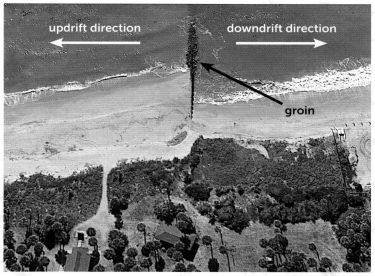

Groin at Edisto Beach State Park. The longshore current when the photo was taken was moving from left to right.

A Little Philosophy

Engineers construct groins when there's an erosion problem. A groin traps sand and widens the beach on the updrift side, which makes homeowners there happy. But on the downdrift side the beach erodes faster, becoming narrow and indented. To keep a beach in front of every house engineers usually build a series of groins (called a groin field).

But some think groins are ugly, and at low tide they present an obstacle that beachcombers have to walk around. They also obstruct the flow of sand from one island to another. The same groins that maintain beaches on one island rob the adjacent island of its sand supply, causing the beaches on that island to erode more quickly. Sometimes piers also have a groin effect and accumulate a little sand on the updrift side.

A jetty looks like a groin but is built next to an inlet. Its purpose is to keep shipping channels from filling in with sand. Like groins, jetties block the longshore transport of sand, starving beaches of sand on the downdrift side (see Activity 10). Piers, jetties and groins are risky places to swim because they sometimes create powerful rip currents (Activity 5).

Are seawalls and groins good or bad for beaches? How you answer this question depends on your philosophical point of view. Some believe it's okay to control nature because nature exists for our benefit. Seawalls, groins and replenishment projects are needed to protect buildings from the encroaching sea. Others believe nature should be left alone. A natural beach is beautiful, and the plants and animals living there do best without human interference. Which of these points of views do you think has greater merit? Why?

Places with Seawalls and Groins

Seawalls and groins can be found on many Carolina beaches. A good example of how groins create erosion indentation is the single groin at the Cape Hatteras Lighthouse. This shows why groins are usually found with others in a groin field.

	Seawalls	Groins
North Carolina		
Bald Head Island	✓	✓
Cape Hatteras Island		✓
Carolina Beach	✓	
Figure Eight Island	✓	✓
Fort Fisher	✓	✓
Ocean Isle Beach	✓	
Pea Island	✓	
South Carolina		
Daufuskie Island		✓
Debidue Island	✓	
Edisto Beach		✓
Folly Beach	✓	✓
Fripps Island	✓	
Hunting Island	✓	
Myrtle Beach	✓	
Pawleys Island		✓
Sea Brook Island	✓	
Sullivan's Island		✓

✓ sandbags

Litter

Water and air, the two essential fluids
on which all life depends, have become global garbage cans.
—Jacques Yves Cousteau
(French oceanographer, explorer and filmmaker)

Nearly every ocean-facing beach on the planet, including those on remote islands hundreds of miles from the nearest city, is littered with disgraceful heaps of trash—bottles, Styrofoam, plastic, aluminum cans, fishing floats, lumber and other assorted discards of our industrial civilization. Most visitors to East Coast beaches are unaware of the magnitude of the problem. That's because trash at popular tourist beaches is picked up, often on a daily basis. It's a sad irony that beaches farthest from urban centers never have their trash picked up and are often the ones with the most garbage.

But litter is more than just an esthetic issue and a hindrance for those wishing to commune with nature. What washes up on a beach can be harmful to humans and deadly to marine life. Plastic garbage in particular is a great evil. More than a million seabirds die every year from ingesting too much plastic. And there are other serious, though less obvious problems associated with plastic.

The purpose of this activity is to determine the origin of garbage on a beach, to learn why litter is harmful to the marine ecosystem and to figure out ways to reduce the flood of garbage washing over the world's beaches.

Determine the Origins of Litter

Beach trash comes from four sources: 1) beach visitors, 2) land trash carried to the sea by rivers, 3) local fishing boats and passing freighters and 4) storms (and tsunamis). The following activity is best done on beaches with few visitors where trash is most abundant.

WHAT YOU NEED A beach with trash, a large kitchen garbage bag, a pencil and notebook and the energy to stroll long distances.

WHAT TO DO Walk along the beach and bag any trash you see. Empty your bag when it's full and divide the litter into four piles: 1) local trash, 2) land trash, 3) trash from boats and 4) trash from storms.

To determine litter origin it helps to read the labels. It's not uncommon to find writing in Chinese, Spanish, Russian, Japanese, Arabic and a host of other non-English languages. Such trash is thrown overboard from foreign vessels (along with debris tossed by English-speaking crews). Other seaborne trash includes fishing discards, boat hatches and pieces of shipwrecks. Surfing wax, sunglasses, sunscreen bottles and the like are probably the work of local litterbugs.

What was the most common kind of trash that you found at your beach? What percentage of the trash do you think is local? What percentage came from sea or from land sources? (See Activity 23 for help with percentages.)

Record your conclusions in your notebook. Include a list of the different kinds of litter you found. Then bag your trash and bring it to a proper waste facility. Make sure to separate recyclables and any hazardous waste.

Islands Near the Stream

According to the EPA, 80 percent of beach garbage comes from land and 20 percent comes from the sea. Land trash blown by wind or washed through storm drains into rivers eventually finds its way to the coast. There it gets carried by wind, tide and longshore currents to beaches. Trash from the sea comes from freighters, container ships, fishing vessels, offshore oil rigs and recreational boats.

   The origin of garbage on beaches varies somewhat according to location. Beaches on the Outer Banks, for example, have a lot of seaborne debris because the beaches are far from the mainland and close to the Gulf Stream. The Gulf Stream is a warm ocean current that flows north from the Caribbean, approaching to within 12–15 miles of the Outer Banks. In addition to bringing coconuts, tropical seeds, bamboo and exotic hardwoods like mahogany to the Outer Banks, the Gulf Stream also delivers garbage of all kinds, some of it from as far away as Cuba and other Caribbean islands.

Plastic trash is a common sight in beach wrack.

Problems with Plastic

Human beings lived without plastic for hundreds of thousands of years. Now it's hard to imagine life without it. Plastic is a material of great convenience. It is also a disaster for marine life. Birds and turtles eat plastic bags, mistaking them for food like jellyfish. The plastic clogs their intestines. The animals starve. Seals, dolphins, birds, fish and turtles get entangled in plastic nets and plastic rings. They suffocate or drown. Plastic absorbs toxic chemicals like PCBs and DDT. Fish ingest plastic and transfer the poisons up the food chain to humans.

Plastic Pollution Solutions

You can limit the impact of plastic on marine life if you:

- Reduce your use of plastic, especially plastic bags.
- Recycle everything.
- Never litter.
- Clean up your beach.
- Don't put trash in a storm drain.
- Tell others about the problems with plastic.

Dead albatross with plastic debris in its stomach.

Beach Driving

Wilderness needs no defense—it only needs defenders.
—Edward Abbey (writer, environmentalist and reputed member
of the Monkey Wrench Gang)

People all over the world have enjoyed the pleasures of driving (and racing) on beaches since the invention of the automobile. To bounce along an empty stretch of sand with surf thundering in the background, the smell of salt in the air, a line of dunes on one side, the boundless blue sea on the other can be a thrilling experience. It's also a quick way to get to your favorite surfing or fishing spot.

But beach driving hurts the plants and animals living there. It also affects the esthetic experience of nondrivers. Such a fragile and unique part of nature should be handled with care, so it's important to understand what happens to beaches when you drive on them.

WHAT YOU NEED A beach where driving is allowed.

WHAT TO DO Walk around a beach where tire tracks can be found. You must check locally because many communities allow beach driving only during the off-tourist season and sometimes the rules change from year to year. Look at several beaches if possible, including those where driving is not allowed. Fishing piers offer an excellent view of driving patterns.

Look closely at the tire tracks. Can you see any signs of damage done to the beach? Consider how driving might affect the critters, both large and small, living in the sand. How does it affect dune plants? Should driving be allowed on all beaches, some beaches or no beaches? What's wrong with driving on a beach?

Driving in the dunes kills the plants that keep dunes in place.

Deep tire tracks make walking difficult for the elderly and the very young.

Cars chase nesting birds away and even destroy their nests.

Driving compacts sand in the intertidal zone, which may harm the meiofauna living there.

For those who enjoy the quiet solitude of nature cars on a beach can be a major irritation.

Turtle hatchlings sometimes get trapped in tire ruts and never make it to the sea. They can also mistake headlights and other artificial lights for natural moonlight and crawl in the opposite direction from the sea.

Ghost crabs and other beach critters are crushed under the tires of cars.

Beach driving breaks shells that otherwise might be of interest to the beachcomber.

Ocean Acidification

In an age when man has forgotten his origins . . .
water along with other resources has become the victim of his indifference.
—from Rachel Carson, *Silent Spring* (the book that launched
the environmental movement)

A Carolina beach is composed of quartz and other minerals mixed with
varying amounts of shell fragments. Cape Hatteras seems to be the magic
boundary between high and low concentrations of shells. Those beaches
north of Cape Hatteras (all the way up to Maine) usually have around 1
percent shell material. Beaches south of Cape Hatteras have 10 percent or
more of shell material. Of course, these percentages can be thrown off if
the beach has been replenished (see Activity 31).

Sometimes it's hard to see the shell fragments in beach sand. They are
either too small or too few to tell apart from mineral grains. But there's
a quick way you can confirm their presence, even if you can't see them.
Seashells are made of calcium carbonate. Calcium carbonate reacts
with acid. Drop some acid on beach sand and watch the sand fizz and
bubble as the acid reacts
chemically with any shells
that may be present. As
we shall see, this has
important consequences
for the health of the world's
oceans, especially for those
friends of Mr. Crab whose
bodies contain calcium
carbonate.

The shells of crabs are made of chitin
(the same material as an insect's outer
body) mixed with calcium carbonate.

Testing for Carbonates with Vinegar

WHAT YOU NEED A jar, a bottle of vinegar, a shell and beach sand.

WHAT TO DO You can do this activity on the beach or at home. Place a small shell (or a shell fragment) in a jar of vinegar. The vinegar should be fresh (recently purchased at the supermarket). What happens to the shell after a minute or two? What changes do you see? Why? Now replace the shell in the jar of vinegar with a handful of beach sand. Does vinegar react the same with beach sand as it does with a shell? Why?

Try the following experiment at home. Place a thin shell (like a coquina shell) or shell fragment in a jar of fresh vinegar. Do not put a lid on the jar. Leave the shell in the jar overnight. Come back the next day and see what's left of the shell.

Coquinas are small clams living in the surf zone.

The pH Scale

The pH scale is a measure of how acidic or basic a liquid is (think of a base as the chemical opposite of an acid). The scale goes from 0 to 14. A pH less than 7 is acidic. A pH of 7 is neutral. A pH greater than 7 is basic. Acids and bases occur in nature. For example, some of the snails discussed in Activity 30 use sulfuric acid to penetrate clam shells.

Vinegar is about 5 percent acetic acid (pH 3) by volume. In your experiment you should have seen bubbles forming on the shells and in the sand. The bubbles are carbon dioxide (CO_2) released by the chemical reaction between acetic acid in the vinegar and calcium carbonate in the shells. Leave a shell long enough in fresh vinegar and it will dissolve completely away.

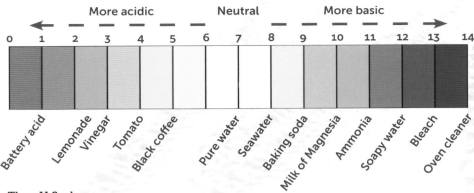

The pH Scale

Ocean Acidification

Scientists have recently noticed that the oceans are becoming more and more acidic. This is due to increasing levels of **carbonic acid**. Carbonic acid forms when carbon dioxide (CO_2) dissolves in seawater. A lot of the carbon dioxide entering the sea comes from the emissions of cars, planes, coal-fired power plants and factories. As CO_2 levels increase, more carbonic acid is formed and the ocean pH decreases (becomes more acidic). What happened to your shells in the vinegar experiment is what's happening to some marine organisms. Corals and the shells of clams, oysters, snails and other marine animals are thinning as their calcium carbonate reacts with carbonic acid in seawater. To learn more about ocean acidification do a search on the NOAA (National Oceanic and Atmospheric Administration) website: www.noaa.gov.

EXPLORE YOUR WORLD
Compare the reactions of vinegar on beach sand at different locations on the beach, including the lower and upper beach and the dunes. Put some black sand (heavy mineral sand, Activity 13) in vinegar. Is there a reaction? Why or why not?

CHAPTER

8

Rainy Day Activities

When you can do nothing, what can you do?

—Zen koan (a paradoxical statement)

Your family was planning to spend the day lazing on beach towels under the warm summer sun, building sand castles and cooling off in the surf. But the forecast is for a cold, wet, miserable day of unrelenting rain. Does that mean you must stay trapped inside, forced to watch stale movies and play computer games all day?

Do not despair. We've put together a list of fun science-related activities that can be done even when the weather doesn't cooperate. Activity 36 explains how to use a microscope to see the miniature cosmos hidden in a single drop of seawater. Activity 37 describes how to boil seawater and use a hydrometer to measure the salt content of ocean water. The next activity lists some of the many science museums, aquariums and archeological/ historical sites located on or near the coast (places you might want to investigate even on fair-weather days). Activity 39 explains how to use Google Earth to learn more about beaches, barrier islands, ocean basins and features on the coastal plain.

Let the good rains fall . . .

Plankton

If you're shopping for a home entertainment system
you can't do better than a good dissecting microscope.
—Jack Longino, "The Astonishing Ant Man,"
on National Public Radio

Plankton are organisms that drift in the sea or in freshwater lakes or rivers. They vary in size from the microscopic to those as big as a man (for example, jellyfish). Plankton that get energy from sunlight are called phytoplankton. Phytoplankton are usually green or blue-green in color. Plankton that feed on other plankton are called zooplankton. A single drop of seawater may contain hundreds or even thousands of plankton.

Most sea animals spend at least part of their life cycle as plankton. Some animals, like fish or barnacles, have a temporary plankton stage. Other organisms are plankton throughout their lives. Zooplankton are usually larger in size than phytoplankton and are able to move around (sometimes rather quickly when viewed with a microscope).

What do plankton look like? Why are such tiny critters critical to the earth? What happens if you swallow seawater full of plankton?

WHAT YOU NEED A microscope (can be purchased at some bookstores and toy stores or online), a notebook and a lidded jar.

WHAT TO DO Fill a jar with seawater. Place a drop of seawater on a microscope slide. Examine the seawater under low magnification. You will have to refocus your microscope since the plankton will be at different levels in the water. Can you distinguish zooplankton from phytoplankton? How many different kinds of plankton can you recognize? Draw a sketch of what you see.

Diatoms are microscopic phytoplankton with a silica shell (something like algae inside a quartz crystal). Some are very beautiful.

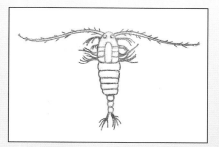

Copepods (.04 inches long or smaller) are the most common kind of zooplankton. With a low-power microscope you can easily see them swimming.

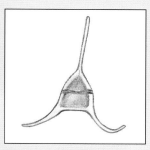

Dinoflagellates are photosynthetic, but some also feed on other plankton. Some are known for their bioluminescence.

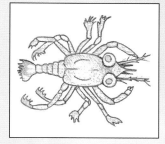

Crab larva. About .1 inch long.

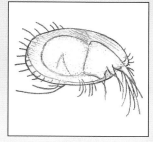

Ostracod (seed shrimp). .04 inches long.

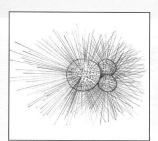

Foraminifera shells are a main component of limestone.

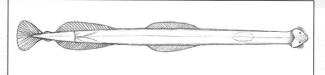

Arrow worms (.08–4 inches long) are predators that feed on copepods and other zooplankton.

Size Doesn't Matter

Plankton are the basis for all life in the ocean and as it turns out, for life on land as well. Their importance is out of proportion to their size.

Over half the oxygen we breathe comes from phytoplankton. Like land plants, phytoplankton use sunlight and carbon dioxide to make energy for growth. Oxygen is given off as a by-product. Even though phytoplankton are small, they are so numerous and so widespread over the vast ocean that their output of oxygen is greater than that of land plants.

Plankton are the base of the ocean's food web. They feed everything from microscopic zooplankton to fish, multi-ton whales and humans. Without plankton, small fish would have no food. Larger fish that feed on small fish would starve. Top predators like bluefish, tuna and sharks would have nothing to eat.

Zooplankton help rid the atmosphere of excess carbon dioxide, and this may slow down global warming. Carbon dioxide (CO_2) from the atmosphere dissolves in the sea. A molecule of carbon dioxide has one carbon atom and two oxygen atoms. Through photosynthesis phytoplankton incorporate the carbon into their bodies and give off oxygen. Zooplankton feed on phytoplankton. When zooplankton die, their bodies fall to the sea floor, taking carbon with them. In this way CO_2 is removed from the atmosphere (the process is known as a carbon sink). Carbon dioxide in the atmosphere traps sunlight, which heats up the air, which melts glaciers, which causes sea level to rise. The plankton carbon sink helps slow the warming of the atmosphere. Have you thanked a plankton today?

Sea foam is made when waves whip and churn organic matter into air bubbles. The organic matter includes proteins and fats from diatoms, algae, dinoflagellates and other sources. Insects, meiofauna and long-wristed hermit crabs feed on foam.

Red Tides

Sometimes phytoplankton reproduce too quickly and become so abundant that the ocean turns green, brown or even red. These are called algal blooms. When certain kinds of dinoflagellates are involved, the water may turn pink or red (hence the name "red tide").

Some red tides are poisonous and can kill large numbers of fish, birds, sea mammals and mollusks. People can get sick from eating seafood contaminated by red tides.

Draw a Food Web

A food web is a visual representation of who eats what. Life can be divided into two basic groups: 1) producers like plants and phytoplankton that get energy from the sun and 2) consumers that get energy by eating other life. Sometimes drawing a food web clarifies the connections between producers and consumers.

WHAT YOU NEED A sketchbook/notebook and a pencil.

WHAT TO DO Use examples in this book (or do your own research) to draw a food web like the example below. How are other marine animals, birds, crabs and humans connected? Draw another food web for the beach.

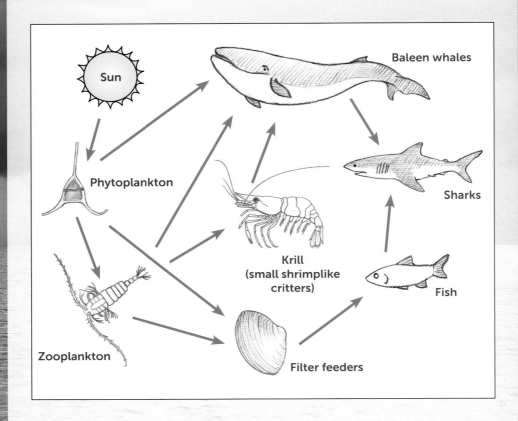

Saltwater

Water, water every where,
Nor any drop to drink.
—Samuel Taylor Coleridge,
"The Rime of the Ancient Mariner"

If you've ever swum in the ocean, you've probably swallowed seawater. The taste of the water combined with your stinging eyes tells you the sea is salty. In fact, we all swallow a little seawater everyday! That's because the salt we sprinkle on our food came from the sea, either from deposits from the distant geologic past or from salt works in hot, dry coastal zones where people make salt by evaporating seawater.

As rain seeps through rocks and soil, it dissolves some of the minerals. Rivers have been bringing these dissolved minerals (also known as salts) to the ocean for millions of years. In addition to rivers, undersea volcanoes also add salts.

There are many kinds of salts in the sea. Some are made of the elements sodium, chlorine, magnesium, potassium, calcium and, to a lesser extent, silica. Rivers are constantly dumping these elements into the sea, yet the amount of salt in the world's oceans remains constant. That's because these elements are removed from the sea in various ways. Some are deposited on the sea floor as sediments. Some are extracted from the water by clams, crabs, diatoms, corals and other marine animals to build their shells. The amount of salt removed every year is the same amount that is added every year by rivers and volcanoes. Thus the sea never gets saltier.

Prove the Sea Is Salty

It is *not* a good idea to taste seawater to prove the sea is salty. Untold numbers of bacteria and other microbes swim in the sea. Some of them will make you very sick. Many more have yet to be discovered. A safe way to "taste" the salty sea is to mix a teaspoon of table salt in a glass of tap water and then take a sip. That glass has about the same salt content as seawater. You can also do the following experiment.

WHAT YOU NEED A bottle of seawater, a pot or frying pan, a magnifying glass and a kitchen scale (optional; can be purchased at most large hardware stores).

WHAT TO DO Pour some seawater into the pot or frying pan. Boil the water until it has completely evaporated. You should see a residue of salt on the bottom of the pan. After the pan has cooled, taste the residue to verify that it is in fact salt (boiling kills the bacteria). Examine the salt under a magnifying glass. Can you see any crystals? Compare the crystals to table salt. Are the crystals the same or different?

Repeat this experiment by boiling tap water or, better yet, river water that you obtained on the way to the coast. How much salt remains in the pan?

A more advanced version of this activity requires using a kitchen scale to weigh the water before boiling and the salt residue after boiling. If you are careful in your measurements, you will find the concentration of salt in seawater to be roughly 3.5 percent. In other words, 100 pounds of seawater holds on average about 3.5 pounds of salt.

Salinity

The most common salt in seawater is sodium chloride or table salt. Salinity is the measure of how salty the water is. Salinity is often measured in parts per thousand. The average salinity for ocean water is around 35 parts per thousand (written as 35 ppt). Near river mouths the salinity will be lower (30 ppt or less). In a hot, dry climate with high evaporation rates, such as in the Red Sea (between Egypt and Saudi Arabia), the salinity will be higher (about 40 ppt).

Fun with a Hydrometer

A hydrometer is a convenient instrument for measuring salinity. People with saltwater fish tanks use hydrometers to make sure their tanks have the proper salinity. Fishermen use hydrometers to predict what kinds of fish might be caught at a particular location.

WHAT YOU NEED A hydrometer (sold at pet stores and some tackle shops).

WHAT TO DO First, check if your hydrometer is working correctly. Fill it to the correct level with tap water. It should report a reading close to zero parts per thousand. Next, measure the salinity in the water at the beach. How close is your reading to the oceanwide average of 35 ppt? Is there a difference in salinity at different spots on a beach? What about the back side versus the ocean side of a **barrier island**? What about the water in an **estuary**? What happens to the salt content of seawater after a heavy rain? How do you think salinity affects marine life?

The Salt of the Earth

Ocean salinity should be more or less the same along a particular beach but may vary somewhat in response to local conditions. For example, a nearby source of freshwater such as a river or even groundwater seeping through the sand will decrease salinity. Heavy rains will likely make surface waters less salty, while periods without rain will make surface waters more salty. Ocean currents and winds can also affect salinity by mixing waters of different salt concentrations.

Differences in salinity can have a profound effect on the distribution of marine life. The body fluids of marine animals exist in a delicate balance with the salt content of the surrounding seawater. If the salinity is too high, water will flow out from the cells of an organism into the sea and the organism will dehydrate. If the salinity gets too low, water will flow from the sea into the organism's cells, causing the cells to burst. That's why river fish that evolved to live in freshwater don't live in the sea and why ocean fish usually can't swim up a river. There are exceptions. Salmon and eels can live in both freshwater and saltwater. Another exception is the bull shark. Bull sharks have been caught hundreds of miles up the Mississippi River as far north as Illinois and, according to National Geographic, thousands of miles up the Amazon River.

A hydrometer is a device for measuring salinity.

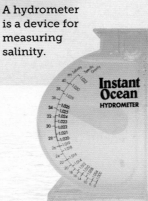

Museum Visit

We live in a world of reproductions—the museums are real.
It's a way to get away from the overload of digital technology.

—Thomas P. Campbell (museum director)

Listed by location from north to south, the following are museums, aquariums, archaeological and historical sites along the coast.

North Carolina

CURRITUCK BEACH LIGHTHOUSE
1101 Corolla Village Road, Corolla, NC 27927
(252) 453-4939
www.currituckbeachlight.com
Visitors allowed to climb the 162-foot-tall lighthouse.

OUTER BANKS CENTER FOR WILDLIFE EDUCATION
1160 Village Lane, Heritage Park, Corolla, NC 27927
(252) 453-0221
www.ncwildlife.org/Learning/EducationCenters/OuterBanks.aspx
Aquarium and exhibits on hunting, fishing and ecology for the area.

NORTH CAROLINA AQUARIUM AT ROANOKE ISLAND
374 Airport Road, Manteo, NC 27954
(252) 475-2300
www.ncaquariums.com/roanoke-island

FORT RALEIGH NATIONAL HISTORIC SITE
1401 National Park Drive, Manteo, NC 27954
(252) 473-5772
www.nps.gov/fora
History of the first English colony in the New World. Also site of the outdoor theater production of *The Lost Colony*.

ROANOKE ISLAND FESTIVAL PARK

Downtown Manteo

www.roanokeisland.com

Elizabeth II (replica of a sixteenth-century sailing vessel), history museum.

AURORA FOSSIL MUSEUM

400 Main Street, Aurora, NC 27806

(252) 322-4238

www.aurorafossilmuseum.com

Not easy to get to but worth the visit. Great collection of Miocene Age fossils from the nearby phosphate mine. Kids can look for fossils for free in the outdoor fossil pit.

CAPE HATTERAS LIGHTHOUSE

46368 Old Lighthouse Road, Buxton, NC 27920

(252) 473-2111

www.nps.gov/caha

GRAVEYARD OF THE ATLANTIC MUSEUM

59200 Museum Drive, Hatteras, NC 27943

(252) 986-2995

www.ncmaritimemuseums.com/graveyard-of-the-atlantic

History of shipwrecks and piracy on the Outer Banks.

FORT MACON STATE PARK

2303 East Fort Macon Road, Atlantic Beach, NC 28512

(252) 726-3775

www.ncparks.gov/Visit/parks/foma/activities.php

Civil War fort and museum.

NORTH CAROLINA MARITIME MUSEUM IN BEAUFORT

315 Front Street, Beaufort, NC 28516

(252) 728-7317

www.ncmaritimemuseums.com/beaufort/

Artifacts from Blackbeard's ship *Queen Anne's Revenge.* Boatbuilding, whaling and local fisheries.

NORTH CAROLINA AQUARIUM AT PINE KNOLL SHOALS

1 Roosevelt Boulevard, Pine Knoll Shores, NC 28512

(252) 247-4003

www.ncaquariums.com/pine-knoll-shores

CAPE FEAR MUSEUM OF HISTORY & SCIENCE

814 Market Street, Wilmington, NC 28401

(910) 798-4370

www.visitnc.com/listing/cape-fear-museum-of-history-science-1

History and natural history of the area, including a giant sloth

NORTH CAROLINA AQUARIUM AT FORT FISHER

900 Loggerhead Road, Kure Beach, NC 28449

(910) 458-8257

www.ncaquariums.com/fort-fisher

FORT FISHER

1610 Fort Fisher Boulevard

South Kure Beach, NC 28449

(910) 458-5538

www.nchistoricsites.org/fisher/

Earthworks-style Civil War fort and a museum.

NORTH CAROLINA MARITIME MUSEUM AT SOUTHPORT

204 E. Moore Street, Southport, NC 28461

(910) 457-0003

www.ncmaritimemuseums.com/southport/

Maritime history, ship models, hurricanes, Civil War.

MUSEUM OF COASTAL CAROLINA AND INGRAM PLANETARIUM

Two museums four miles apart:

MUSEUM: 21 East 2nd Street, Ocean Isle Beach, NC 28468

(910) 579-1016

PLANETARIUM: 7625 High Market Street, Sunset Beach, NC 28468

(910) 575-0033

www.museumplanetarium.org

South Carolina

SOUTH CAROLINA CIVIL WAR MUSEUM

857 US-17 Bypass, Myrtle Beach, SC 29577

(843) 293-3377

www.mbisr.com/sccivilwarmuseum.html

SOUTH CAROLINA MARITIME MUSEUM

729 Front Street, Georgetown, SC 29440

(843) 520-0111

www.scmaritimemuseum.org/

THE VILLAGE MUSEUM

401 Pinckney Street, McClellanville, SC 29458

(843) 887-3030

www.villagemuseum.com

Small history museum in picturesque McClellanville.

FORT MOULTRIE

1214 Middle Street, Sullivan's Island, SC 29482

(843) 883-3123

www.nps.gov/nr/travel/charleston/sum.htm

Visitor's Center and Civil War fort. Accessible by car.

PATRIOTS POINT NAVAL AND MARITIME MUSEUM

40 Patriots Point Road, Mount Pleasant, SC 29464

(843) 884-2727

www.patriotspoint.org

USS *Yorktown* and other vessels.

FORT SUMTER

Charleston Harbor

(843) 883-3123

www.nps.gov/fosu/index.htm

Where the Civil War began. Reachable by ferry.

CHARLESTON MUSEUM

360 Meeting Street, Charleston, SC 29403

(843) 722-2996

www.charlestonmuseum.org

History of Charleston, natural history, Low Country fossils.

CHARLESTON AQUARIUM

100 Aquarium Wharf, Charleston, SC 29401

(843) 720-1990

scaquarium.org

MACE BROWN MUSEUM OF NATURAL HISTORY

School of Sciences and Mathematics Building, College of Charleston, 202 Calhoun Street, Charleston, SC 29401

(843) 953-5589

www.ssm.cofc.edu/natural-history-museum/index.php

Fantastic fossil collection with specimens dating back hundreds of millions of years.

BEAUFORT HISTORY MUSEUM

713 Craven Street, Beaufort, SC 29902

(843) 379-3079

www.beauforthistorymuseum.com

History and natural history of the area.

EDISTO ISLAND MUSEUM

8123 Chisolm Plantation Road, Edisto Island, SC 29438

(843) 869-1954

www.edistomuseum.org

Edisto Island history.

HUNTING ISLAND LIGHTHOUSE

2555 Sea Island Parkway, St. Helena Island, SC 29920

(843) 838-2011

www.huntingisland.com/lighthouse.htm

Visitors are invited to test their fortitude by climbing the 136-foot-tall lighthouse.

COASTAL DISCOVERY MUSEUM

70 Honey Horn Drive, North End, Hilton Head, SC 29926

(843) 689-6767

www.coastaldiscovery.org

Garden, educational boardwalks, nature exhibits, historic buildings and a 400-year-old cedar tree.

Google Earth

What's the use of a fine house
if you haven't got a tolerable planet to put it on?
—Henry David Thoreau
(American philosopher, writer and environmentalist)

Google Earth is a view of the earth based on satellite imagery that allows you to "fly" for free around the world without the hassle of boarding lines and metal detectors.

WHAT YOU NEED A digital device with internet connection and Google Earth (or Maps for Macintosh).

WHAT TO DO Find a view of the North Atlantic that allows you to see both the west coast of Africa and the east coast of North America. Find the Mid-Atlantic Ridge—the north-south-trending ridge in the middle of the Atlantic Ocean. The Mid-Atlantic Ridge is where two plates are splitting apart. North and South America are on one plate. Africa and Eurasia are on the other. More than 200 million years ago the Carolinas were attached to Africa where the Mid-Atlantic Ridge now stands today. Then the supercontinent Pangaea began to break up and the continents drifted apart. They've been drifting ever since.

Zoom in to where you can see the coast from Savannah, Georgia, to Virginia Beach, Virginia. The dark blue line just east of Charleston, South Carolina, is the edge of the continental shelf. About 18,000 years ago this was where the shoreline stood.

Which of the Carolinas has the wider continental shelf? How does the width of the continental shelf affect tidal range and wave height?

Features of the Coastal Plain

Note the change in color that marks the boundary between the **coastal plain** and the **Piedmont**. The boundary is known as the fall line. About 80 million years ago the fall line was a shoreline fringed with river deltas (like the Mississippi Delta today). The sands from those ancient deltas were later reworked by wind to form the Sand Hills.

Examine the area west of Wilmington, North Carolina. You should see hundreds of lakes scattered throughout the coastal plain. Many have been filled in to make farmland, but their outlines are still visible. These are the Carolina bays. Once thought to be meteor impact craters, scientists now believe the bays may have been created by the wind.

How far to the north and south do the Carolina bays extend?

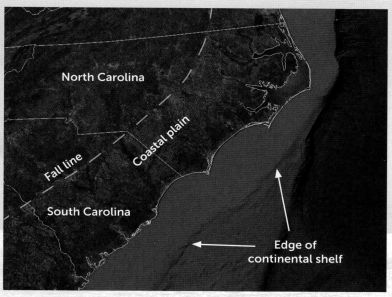

Google map showing edge of continental shelf and approximate location of the fall line.

Barrier Islands

Look at several barrier islands. How many of the following natural and human-made features can you identify?

Maritime forests	Dunes
Tidal deltas	Overwash fans
Swales	Signs of erosion
Salt marsh	Spits
Rip currents	Sandbars
Estuaries	Heavy minerals (black sand)
Non-barrier island shorelines	Artificial (bulldozed) dunes
Navigation channels	Seawalls
Sand bags	Groins
Longshore current direction	Jetties
Replenished beaches	Trees on the beach
Unreplenished beaches	Ripple marks
Inlets	Piers

Compare barrier islands in North Carolina to those in South Carolina. Note the features they share in common. Note also their important differences.

What do the barrier islands in South Carolina have in common with those in North Carolina? How do they differ in shape, width, length and distance from the mainland? How do you explain the differences?

EXPLORE YOUR WORLD

There are more than 2,200 barrier islands around the world. How many can you find using Google Earth?

History

Google Earth has become a useful tool for geologists, historians and archaeologists who interpret the earth's surface to make sense of the past. One entertaining feature of the software is the virtual walking tour it offers for famous landmarks.

GHOST TOWNS

Native American villages, European colonies, fishing villages and resort communities once thrived on the barrier islands of the Carolinas, only to be later abandoned because of war, storms or economic hard times. The best preserved of these ghost towns is the town of Portsmouth on Portsmouth Island in the Outer Banks.

THE FORTUNES OF WAR

Not much is left of the sixteenth-century Spanish forts built in North and South Carolina. But plenty of other forts, some dating from the Revolution, can be explored with Google Earth. Take a virtual tour of Fort Sumter, Fort Moultrie and Fort Macon.

Why were these forts built where they were built? How were the English ultimately able to colonize the Carolinas even though the French and Spanish arrived here first?

SHIPS

Take a virtual walking tour on the flight decks of the USS *Yorktown* in Charleston Harbor and the USS *North Carolina* in Wilmington. Explore some of the many shipwrecks (including U-boats) lining the shoreline.

LIGHTHOUSES

Discover the many lighthouses in North and South Carolina. How many of the 17 still-standing structures can you find?

9

Seeing the Unseen

A man sees what he wants to see and disregards the rest.
—Simon and Garfunkel (American folk/rock duo)

Three people followed a path through the dunes down to the beach. The first was a botanist, the second was a geologist and the third was a surfer. With scarcely a glance at the beach the botanist inspected a clump of grass. "Ah, sea oats," she said. "Very important for dune ecology." Paying no heed to what the botanist was doing, the geologist stooped to examine beach sand under a hand lens. "Aha!" he exclaimed. "Quartz with a scattering of heavy minerals." The surfer, scanning the horizon intensely, ignored both scientists and bolted straight for the water. "Surf's up!" he shouted.

We see only what our minds are trained to see. The rest is a blur on the edge of our peripheral awareness. The purpose of a book like this is not just to educate and inspire, but also to encourage people to "see" the world around them, especially noting those parts of nature hidden, overlooked or hitherto deemed unworthy of consideration. After all, observation (especially of things people haven't noticed before) followed by inquiry and experiment is how science advances.

The two activities in this chapter require keen powers of observation. Activity 40 lists interesting and unusual phenomena seen at a beach only after sunset. Activity 41 explains how you can create your own beach science activity, either by expanding on a previous activity or by designing your own.

Night

With beauty before me, I walk
With beauty behind me, I walk
With beauty all around me, I walk,
In beauty, it is finished.
—Navajo chant

By day the beach is a familiar land of wave sparkle and sun-fired sand, gulls lazing on the breeze, sea oats quietly turning sunshine into sugar. Then the sun sets. Plants shut down their solar factories. The wind reverses direction. Creatures rarely seen by day creep across the sand in search of food and nesting sites. Strange lights glow in the water. The beach at night . . . the unfamiliar country. If your timing is good, a beach after dark will reward you with unusual, even memorable life experiences.

The Green Flash

Just before sunrise and sunset, if atmospheric conditions are ideal, the sun briefly turns a bright, emerald green. This rare phenomenon is known as the green flash. Lasting but a second or two, the green flash happens because the atmosphere refracts (bends) sunlight, separating the light into different colors (much like a prism). In the Carolinas a good place to view this unusual event is on the Outer Banks, where you can watch the sunrise on the open ocean and the sunset over the sound.

WHAT YOU NEED A clear sky, patience and a view of the horizon.

WHAT TO DO Watch the sky a few seconds before sunrise or sunset. *Do not* look directly at the sun as it can damage your eyes. If weather and sky cooperate, you may be one of the fortunate few to witness a green flash.

Ghost Hunter

Ghost crabs (*Ocypode quadrata*) are the sprinters of the crab world. More active at night than by day, ghost crabs are so well camouflaged that most people only see them as a blur of movement as they zip sideways, backward or forward across the sand. They live in burrows up to four feet deep, venturing out to get food or to return to the water long enough to wet their gills. Sometimes you can observe them in their burrows after dark.

WHAT YOU NEED A flashlight, curiosity and a brave heart.

WHAT TO DO Walk along the upper beach with a flashlight. Look for holes marking the entrances to ghost crab burrows. Point your flashlight down a hole and look into a burrow. What do you see? What is unusual about the claws of a ghost crab? You might have to visit several burrows as the occupants may be out scavenging for food. Ghost crabs are a widespread group consisting of 22 species found on sandy beaches all over the world. They are opportunistic feeders (they eat what's available) that eat carrion in the wrack, mollusks, turtle eggs and newly hatched turtles.

Try shining your flashlight toward the surf. Sometimes you can spot seabirds fishing in the swash or dolphins swimming in the surf. Point your flashlight toward the area behind the dunes and into the maritime forest (if any forest is present). Look for the glow of eyes. Learn to identify an animal by its eye shine. Rabbit eyes shine red. A deer's eyes are a very bright amber color and stare continually at a light source. Coyote and bobcat eyes shine green-gold, while raccoon eyes are yellow. Spiders, owls, opossums, skunks, foxes and even fish have eye shine. *But be careful!* A large pair of red eyes low to the ground may be those of an alligator! Alligators don't like saltwater but occasionally show up on beaches in both Carolinas.

Things that Glow in the Dark

Some life forms use a chemical reaction to emit their own light. This is known as bioluminescence. In the sea bioluminescent organisms include some kinds of bacteria, algae, squid, fish, comb jellies, sea stars, sharks, sea jellies (jelly fish) and single-celled plankton called dinoflagellates. Fireflies use bioluminescence to find mates. Marine organisms use it to distract or warn predators or to lure prey. Look carefully and you might see bioluminescence in harbors and on beaches in the Carolinas.

WHAT YOU NEED A moonless night and a little luck.

WHAT TO DO Walk along the edge of the water. Look in the ocean water and on the sand for the blue-green sparkle of bioluminescent organisms. Sometimes an otherworldly glow appears on wet beach sand after you walk or jump on the sand. Other times you can fire up bioluminescent sparkles by dragging your hand through seawater, leaving a watery trail of blue-green diamonds.

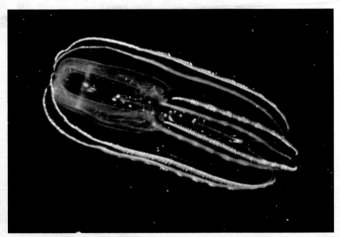

Bolinopsis infundibulum is a comb jelly. Comb jellies are similar to jelly fish (sea jellies) but belong to a different group. Some are bioluminescent.

Nesting Sea Turtles

All seven species of the world's sea turtles have been around since the dinosaurs. All are declining in numbers. Thousands die every year from fishing nets or ingesting plastic or are eaten by people. They are also threatened by marine pollution, beach driving, habitat loss and artificial lighting. Four species nest on the beaches of the Carolinas: the loggerhead, the leatherback, green turtles and the Atlantic (Kemp's) ridley. At night from late April to September turtles crawl from the sea, dig a hole in the upper beach, lay and bury their eggs and then crawl back to the sea. About 45–60 days later the hatchlings make their way to the ocean.

WHAT YOU NEED A beach at night and *no flashlights* or other lights (artificial lights discourage mothers from nesting and confuse hatchlings, causing them to wander inland).

WHAT TO DO Observe female turtles laying their eggs or the hatchlings crawling to the sea. You can do this on your own or join a group tour. Organizations that sponsor turtle nesting tours include Bald Head Island Conservancy (North Carolina), Coastal Discovery Museum (Hilton Head) and Edisto State Park (South Carolina). Many coastal communities have volunteers that protect the eggs from predation (mostly from raccoons but also from ghost crabs, foxes and, unfortunately, humans).

Most sea turtles in the Carolinas are loggerheads.

Is Anybody Out There?

WHAT YOU NEED Binoculars, a star chart and a sense of wonder.

WHAT TO DO Identify major constellations, stars and the Summer Triangle (marked by Deneb, Altair and Vega). Find Polaris (the North Star) by using the "pointers" of the Big Dipper. With binoculars you can see four of Jupiter's moons, the rings of Saturn, star clusters, moon craters and comets.

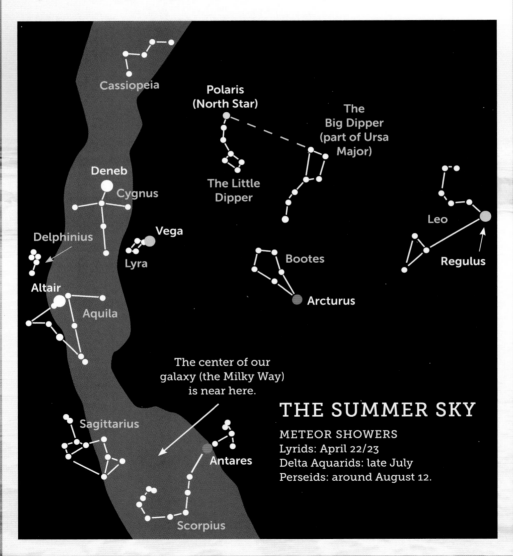

Cassiopeia

Polaris (North Star)

The Big Dipper (part of Ursa Major)

Deneb

Cygnus

The Little Dipper

Delphinius

Vega

Lyra

Altair

Aquila

Bootes

Leo

Regulus

Arcturus

The center of our galaxy (the Milky Way) is near here.

Sagittarius

Antares

THE SUMMER SKY

METEOR SHOWERS
Lyrids: April 22/23
Delta Aquarids: late July
Perseids: around August 12.

Scorpius

The best beaches for stargazing are those farthest from city lights. Test your eyesight by counting the stars in the Pleiades (you should be able to see six or more). Find the Andromeda Galaxy, a faint cloudy patch in the constellation Andromeda. Note the different colors of stars—white, red, yellow, blue. About one-third of all stars are multiple systems (two or more stars rotating around each other). Algol (the "Demon Star") is an eclipsing binary. Every 2.85 days Algol gets dimmer as one star passes behind the other. Look for Sirius, the brightest star in our sky. To find out when the International Space Station is visible visit spotthestation.nasa.gov/.

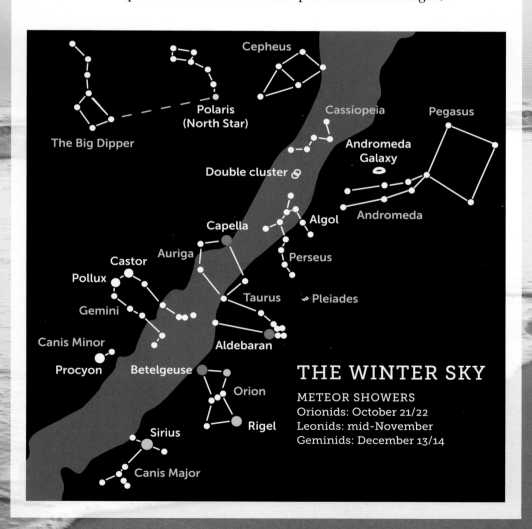

Cepheus

Polaris
(North Star)

Cassiopeia

Pegasus

The Big Dipper

Andromeda
Galaxy

Double cluster

Andromeda

Algol

Capella

Perseus

Auriga

Castor

Taurus

Pleiades

Pollux

Gemini

Canis Minor

Aldebaran

Procyon

Betelgeuse

THE WINTER SKY

Orion

METEOR SHOWERS
Orionids: October 21/22
Leonids: mid-November
Geminids: December 13/14

Rigel

Sirius

Canis Major

Do Your Own Activity

The true sign of intelligence is not knowledge but imagination.

—Albert Einstein (American physicist)

By this time you have done enough activities to be familiar with the process of observation, questioning and experiment. Now it's time for you to design your own activity, either by expanding on a previous activity or by creating your own. Although this book is centered on the narrow band of land between forest and sea, there's no reason why you can't explore other parts of the coastal ecosystem. Salt marshes, swales, maritime forests or estuaries all make excellent subjects of study.

WHAT YOU NEED Imagination, curiosity and a sense of wonder.

WHAT TO DO Walk along the beach (or any other natural landscape) as far as possible from the madding crowds of summer. Look around. Look for things you haven't seen before, sounds you haven't heard, new smells, etc. Ask questions. Why are those birds flying in a line? How are those holes in the sand formed? What is that strange creature tossed ashore by wind and waves? Where does it live? What does it eat? Find answers to your questions by

- *Observation.* Sometimes by observation alone you can understand how nature works.
- *Experiment.* Create a hypothesis explaining your observations. Test the hypothesis with an experiment. Repeat the experiment.
- *Thought experiment* (see Activity 11).
- *Research.* Read science articles, magazines or books. Perhaps others have already found the answers to your questions.

Glossary

adhesion ripple A feature produced when dry sand is blown over a wet surface. The sand sticks to the wet surface to form irregular wartlike features or small irregular ripples.

algal bloom A sudden increase in the amount of algae in nearshore waters, usually caused by an increase in nutrients.

atmospheric pressure The weight of the air pressing against the earth's surface. Atmospheric pressure is also called barometric pressure. Low atmospheric pressure usually means the weather is cloudy, rainy or stormy. High pressure usually means fair, calm weather.

barrier islands Long ridges of sand, backed by a sound (lagoon) and separated from other islands by inlets. Barrier islands, with the exceptions of Myrtle Beach, South Carolina, and Kure Beach, North Carolina, front almost the entire Carolina coast.

barrier island migration Barrier islands move toward the mainland in response to sea-level rise. Migration consists of erosion on the ocean side and widening of the back side by storm overwash. As islands migrate, the mainland shoreline also moves back as it too erodes.

beach replenishment or beach nourishment Taking sand from somewhere else and putting it on a beach to widen the beach for recreation and for storm protection.

bioluminescence Light emitted by life. Fireflies are an example of bioluminescence on land. Dinoflagellates, bacteria and some deep-sea fish are examples in the ocean.

bivalves A group of mollusks that includes clams, oysters, scallops and mussels. Bivalves have shells made of calcium carbonate consisting of two parts (called valves) that open and close on a hinge.

brackish A mix of fresh- and saltwater.

bubbly sand Beach sand that is soft when walked on. Bubbly sand forms when the rising tide pushes air into the sand, creating thousands of holes.

carbonate fraction The portion of beach sand made up of broken-up seashells.

carbonic acid The acid formed when CO_2 is absorbed by seawater. Production of carbonic acid causes ocean acidification, which is one result of climate change.

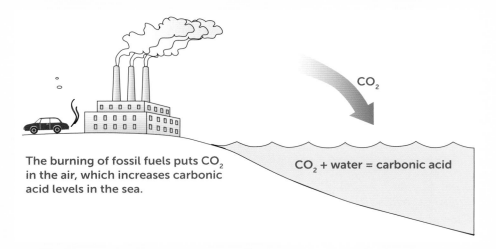

The burning of fossil fuels puts CO_2 in the air, which increases carbonic acid levels in the sea.

CO_2 + water = carbonic acid

coastal plain That part of the east coast (usually more than 100 miles wide) from Long Island to south Florida. The coastal plain is an apron of sedimentary rocks derived from the eroding Appalachian Mountains during the past several million years. Most barrier islands in the world form at the edge of coastal plains.

continental shelf The wide, flat, relatively shallow underwater landmass (shelf) that extends seaward from the shoreline. The continental shelf is about 15 miles wide off Cape Hatteras and about 80 miles wide off Georgia.

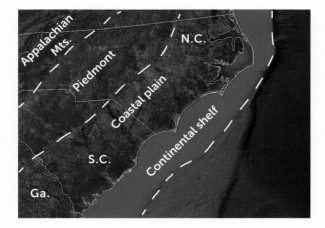

crustaceans A large group of organisms that includes crabs, lobsters, barnacles and shrimp.

current ripple A type of asymmetrical ripple mark (an inch or two high) that forms as water flows on the beach. The steep face of the ripple faces downstream.

dinoflagellate A single-cell organism that is often responsible for bioluminescence in the water column and on beaches.

downdrift The direction on a beach where most sand is transported. In general, on Carolina beaches, south or downcoast is the downdrift direction. Downdrift is similar to downstream in rivers.

dredge A machine that pulls up sediment from the sea floor and pumps it ashore for beach replenishment.

echinoderms A large group of organisms made of calcium carbonate that includes sand dollars, sea stars, sea urchins and sea cucumbers.

ecological niche The role and location of a species in the environment. A niche includes the food an organism eats, how it interacts with other life and how it survives, reproduces and finds shelter.

estuary Where a river meets the sea and freshwater from land mixes with saltwater from tides. In the Carolinas, estuaries are river valleys flooded by sea-level rise. Estuaries and their surrounding wetlands are important feeding and breeding grounds for fish, crabs, birds and other wildlife.

evolution/evolve The process in which plant and animal species change slowly over millions of years to better adapt to their environments. All life evolved from a common ancestor.

feldspar Feldspar, like quartz, is one of the major light minerals in beach sands of the Carolinas. It is difficult to distinguish feldspar from quartz except with a microscope. Typically it is less than 10 percent of the light-mineral fraction.

fetch The distance over open water that wind can blow to create waves. The greater the fetch, the bigger the waves. Since the fetch for the Carolinas is from the shoreline in Morocco to the shoreline of the United States, the potential for big waves is great. Wind speed and duration of the wind also affect wave size.

filter feeders Animals such as clams that filter their food out of seawater.

gastropods A group of coiled mollusks, including snails, whelks, conches and abalone.

global warming The increase in temperature of the earth's atmosphere caused by emissions from cars, factories, planes and other sources.

grain size The diameter of individual particles of sand in beach sediments.

groins Coastal engineering structures consisting of a wall (made of rock, steel, wood or other materials) extending oceanward, perpendicular to the shoreline. They are designed to trap sand and widen the beach on one side, but they cause erosion on their other (downdrift) side.

groundwater Water found beneath the surface of the earth, stored in spaces within soils or between beach sand grains.

Gulf Stream The large, powerful warm ocean current in waters just beyond the continental shelf. The Gulf Stream circles the Atlantic Ocean from Europe to North America and back.

heavy minerals The mineral fraction of beach sand that is heavier than quartz. Usually black but sometimes red or green in color.

ilmenite A black iron-titanium oxide mineral that is an important constituent of heavy minerals in Carolina beach sand. It is partly responsible for the overall black color of heavy-mineral concentrations. In some parts of the world, beaches are mined for their titanium content in ilmenite.

inlet The channel at the ends of all barrier islands through which water flows in and out of lagoons and estuaries behind the island.

intertidal zone The beach surface between the levels of normal high tide and normal low tide.

jetty A wall perpendicular to the beach adjacent to inlets. Intended to keep inlets open for navigation, they often cause serious erosion on adjacent barrier islands.

krill Shrimplike organisms that are an important source of food for whales, seals, fish, squid and other marine organisms.

Labrador current The ocean current that starts in the Arctic Ocean and flows south to Cape Cod and beyond. It cools water as far south as Cape Hatteras.

longshore current The surf-zone current formed by breaking waves that come ashore at an angle. It's responsible for much of the near-shore transport of beach sand. (Also called longshore drift.)

longshore current equation Velocity (speed) equals distance divided by time V=D/T. In this case the distance is 53 feet (about 1/100 of a mile). T is the time in seconds it takes the current to move an orange 53 feet. We could measure current velocity in feet per second, but we want to know the speed in miles per hour. By using the following equation, the units cancel out and longshore current velocity can be calculated in miles per hour (when T is measured in seconds).

$$\frac{53 \text{ feet}}{\text{T seconds}} \times \frac{1 \text{ mile}}{5{,}280 \text{ feet}} \times \frac{3{,}600 \text{ seconds}}{1 \text{ hour}} = \frac{36 \text{ miles}}{\text{T hour}} = \text{velocity of the current in miles per hour}$$

magnetite An iron oxide mineral that, along with ilmenite, is responsible for the black coloration of most heavy-mineral patches on the beach. Magnetite is easily picked up by a magnet.

maritime forest The unique forest commonly found on wide barrier islands behind the dunes. The main trees are cedar and live oak, both of which are capable of withstanding wind and salt spray. In the old days, maritime forests were logged for their live oak timbers used in sailing ships.

mastodon An extinct species that was similar in appearance to mammoths and elephants.

megalodon The largest shark that ever lived. Don't worry, it's extinct!

meiofauna The animals that live between sand grains. They are the all-important bottom of the food chain.

mollusks The large group of invertebrates that includes snails, slugs, bivalves and cephalopods (such as squid and octopus).

neap tide The smallest tidal range at a beach; that is, the smallest vertical difference between high- and low-tide levels, occurring during the first or third quarters of the moon.

nor'easter A general term for a storm approaching the Carolina coast from the northeast, a common event in winter.

offshore wind A wind blowing from land toward the sea.

onshore wind A wind blowing from the sea toward land.

overwash Beach sand transported inland beyond the beach by storm waves. On barrier islands large overwashes in big storms may extend all the way across the island into the sound. This widens the island and is an important element of island migration.

Pangaea The supercontinent that existed millions of years ago before it broke up into North and South America and other land masses.

Pangaea

photosynthesis A chemical process that plants use to create energy from sunlight.

Piedmont The region between the Appalachian Mountains to the west and the coastal plain to the east that is higher in elevation than the coastal plain but lower in elevation than the Appalachians.

plankton Small and microscopic organisms that drift and float in the sea. These are an important food source for many organisms. Plankton that get energy from sunlight are called phytoplankton. Plankton that get energy by eating other plankton are called zooplankton.

Pleistocene The time period that extended from about 2.6 million to 11,000 years ago. This was the time of the Ice Age and the rising and lowering of sea level. See also the geologic time scale on page 216.

plunging breaker A wave that breaks on a moderate beach slope (usually 3–13 degrees). A plunging breaker curls over, forming a barrel or tube of air as it collapses. It is the most forceful type of breaker in terms of generating sand movement on the sea floor and makes spectacular surfing.

pyrite An iron sulfide mineral that, in very fine crystals, is responsible for the black coloration of some shells on the beach

quartz The most common mineral in most beach sands. It is a hard, durable mineral formed of silicon dioxide.

red tide An algal bloom that has a reddish color.

replenished (*see* beach replenishment)

rip current A fast-moving flow of water from the beach through the surf zone—very dangerous to swimmers.

ripple marks Low (usually less than one inch high), long ridges and long troughs that occur in a repetitive pattern. Different patterns are created by different air, water and wave conditions.

rogue wave A rare wave usually in the open ocean, formed by the combining of several different waves that create a much larger wave.

salinity A measure of the amount of salts that are dissolved in seawater. Open ocean salinities are around 35 parts per thousand.

scarp Small cliffs in the sand that reflect a rapid erosion rate or the passage of a recent storm. Most are small, but some vertical scarps may be 10 feet or higher. These are particularly common on replenished beaches. Scarp is also an old term for the eroded remains of old barrier islands found on the coastal plain.

scientific method

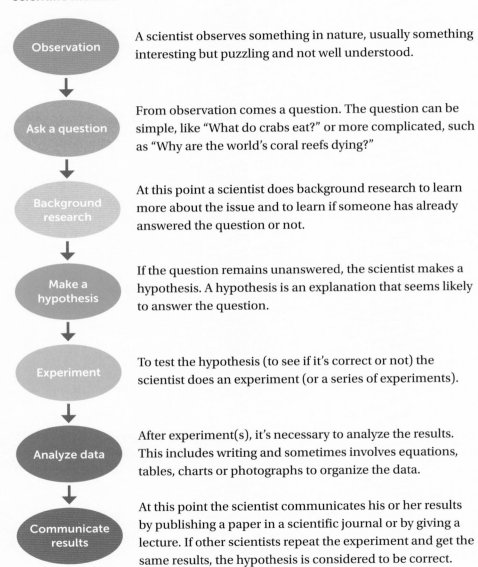

Observation — A scientist observes something in nature, usually something interesting but puzzling and not well understood.

Ask a question — From observation comes a question. The question can be simple, like "What do crabs eat?" or more complicated, such as "Why are the world's coral reefs dying?"

Background research — At this point a scientist does background research to learn more about the issue and to learn if someone has already answered the question or not.

Make a hypothesis — If the question remains unanswered, the scientist makes a hypothesis. A hypothesis is an explanation that seems likely to answer the question.

Experiment — To test the hypothesis (to see if it's correct or not) the scientist does an experiment (or a series of experiments).

Analyze data — After experiment(s), it's necessary to analyze the results. This includes writing and sometimes involves equations, tables, charts or photographs to organize the data.

Communicate results — At this point the scientist communicates his or her results by publishing a paper in a scientific journal or by giving a lecture. If other scientists repeat the experiment and get the same results, the hypothesis is considered to be correct.

sea The choppy, irregular (confused) water surface formed when waves are locally generated and not coming from a distant location.

sea level The average elevation of the water surface of the ocean.

sea oats A common dune plant found throughout the Carolinas south of Cape Hatteras, North Carolina. North of Hatteras, American beach grass takes over.

seawall An engineered wall, made of stone, wood, steel or sandbags, installed on the upper beach parallel to the beach. Seawalls are an effort to prevent retreat of the shoreline and loss of properties at the beach. In the long term, seawalls on eroding beaches always cause the loss of the beach.

shell hash A term used to describe a concentration of broken shell material on a beach.

shoreline retreat A synonym for shoreline erosion.

sorting Refers to the range of grain sizes in beach sand. In well-sorted sand, almost all the grains are the same size. In poorly sorted sand, there is a wide range of particle sizes.

sound A synonym for lagoon or the water body between barrier islands and the mainland.

spartina A common salt-marsh grass in the Carolinas that exists where the plant is flooded by saltwater at least once a day.

spilling breaker A wave that breaks on a relatively flat beach (typically 3 degrees or less slope). The wave crest literally spills over the top of the wave but does not curl like a plunging breaker.

spit A curved or hooklike extension of the beach that is formed by longshore sand transport. Common at the downdrift ends of barrier islands.

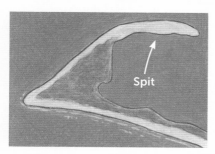

spring tide The highest tidal range at a beach occurring during a full or new moon. Storms occurring during a spring tide are particularly dangerous for beach cottages.

storm surge The rising water level due to water being pushed onshore by storms and also by the low atmospheric pressure of a storm. Such storm surges may cause water to flow miles inland from the sea.

swale A flat area or depression in a dune field sometimes partly filled with water. Also known as a blowout.

swash The thin layer of water that is the final remains of a wave as it rolls up the beach.

swash zone The area over which wave swash is running up and down the beach. The swash zone changes location as the tide rises and falls.

swell Evenly spaced waves with long wavelengths, indicating formation by wind a long distance away.

symbiosis A mutually beneficial relationship between two different organisms.

tidal range (amplitude) The vertical distance between normal high and low tides.

tide gauge A device on piers, docks and sometimes seawalls that records the history of tides at that location. These provide the basis for understanding sea-level rise.

tide table (tide chart) Information concerning the times of high and low tides at a particular location. Tide tables are found on piers, local weather reports on TV, in newspapers and online.

tsunami An immense wave formed as a result of an earthquake, submarine landslide or underwater volcanic eruption. Very rare in the Carolinas.

updrift The direction up the beach in which the least amount of sand is transported—more or less equivalent to upstream in rivers.

water table The level within the earth marking the top of the groundwater.

wave crest The highest point of a wave.

wave direction The direction from which waves are coming.

wave height The vertical distance between the wave crest and the wave trough.

wavelength The distance between two successive wave crests.

wave period The time it takes for a wave crest to pass a given point.

wave ripple A type of ripple mark (formed by waves) that is symmetrical, as opposed to a current ripple mark, which is asymmetrical.

wave train A group of successive waves of the same size and spacing moving in the same direction.

wave trough The low point of a wave.

wet/dry line The point on the beach where the intertidal zone ends and the dry beach begins.

wrack A line of natural and artificial debris on the beach; for example, seaweed, spartina, fishing nets, lumber, driftwood and plastic bottles. Sometimes wrack causes an accumulation of sand to form new dunes. Wrack is an important food source for birds, crabs and other critters.

Geologic Time Scale

		Age (millions of years ago)
Cenozoic	HOLOCENE	0.01 (11,700 years ago)
	PLEISTOCENE	2.6
	PLIOCENE	5.3
	MIOCENE	23
	OLIGOCENE	34
	EOCENE	56
	PALEOCENE	66
Mesozoic	CRETACEOUS	145
	JURASSIC	201
	TRIASSIC	252
Paleozoic	PERMIAN	299
	PENNSYLVANIAN	323
	MISSISSIPPIAN	359
	DEVONIAN	419
	SILURIAN	444
	ORDOVICIAN	485
	CAMBRIAN	541
	PRECAMBRIAN	

The Edge of the Wild

I must go down to the sea again, to the lonely seas and sky,

And all I ask is a tall ship and a star to steer her by.

—John Masefield, "Sea-Fever"

Beach Conservation

Beaches are mysterious and captivatingly beautiful landforms. Sadly, many are in danger of being overrun by human development. Some accept this as an unfortunate but inevitable consequence of progress. Others think we can do better; we can at least preserve some of our beaches in their natural state. If you agree with the latter view and want to do something to preserve beaches, you may choose to join one of the environmental groups listed below. Some groups offer internships, scholarships, summer camps, field trips, lectures and opportunities for volunteer work. The North Carolina Coastal Federation and the Coastal Conservation League are the largest shoreline conservation groups in the Carolinas.

COASTAL CONSERVATION LEAGUE

328 East Bay Street, Charleston, SC 29401

(843) 723-8035

www.coastalconservationleague.org

THE NATURE CONSERVANCY

World Office: 4245 North Fairfax Dr., Suite 100, Arlington, VA 22203-1606

(800) 628-6860

www.nature.org

NORTH CAROLINA COASTAL FEDERATION

3609 NC 24, Newport, NC 28570

(252) 393-8185

www.nccoast.org

NORTH CAROLINA WILDLIFE FEDERATION

1024 Washington Street, Raleigh, NC 27605

(919) 833-1923

www.ncwf.org

THE SANTA AGUILA FOUNDATION

P.O. Box 5006, Santa Barbara, CA 93150

www.coastalcare.org

THE SIERRA CLUB

North Carolina Sierra Club

19 West Hargett Street, Suite 210, Raleigh, NC 27601

(919) 833-8467

South Carolina Sierra Club

1314 Lincoln Street, Suite 211, Columbia, SC 29202

(803) 256-8487

www.sierraclub.org

SOUTH CAROLINA WILDLIFE FEDERATION

215 Pickens Street, Columbia, SC 29201

(803) 256-0670

www.scwf.org

SURFRIDER FOUNDATION

Check the website for local chapters in the Carolinas.

www.surfrider.org

Books

There are thousands of books about beaches; topics include sand, seashells, waves, currents, storms, seawalls and the animals and plants that live in or on the beach. The following might add to the enjoyment of carrying out the activities in this book. The second two books were the sources for many of the activities in this book.

Ashley Oliphant, *Shark Tooth Hunting on the Carolina Coast* (Sarasota, Fla.: Pineapple Press, 2015).

Orrin H. Pilkey, Tracey Monegan Rice, and William J. Neal, *How to Read a North Carolina Beach: Bubble Holes, Barking Sands, and Rippled Runnels* (Chapel Hill: University of North Carolina Press, 2004).

Blair and Dawn Witherington, *Living Beaches of Georgia and the Carolinas* (Sarasota, Fla.: Pineapple Press, 2011).

For young readers we recommend the following:

Steve Parker, *Eyewitness Books—Seashore* (New York: Knopf, 1989).

Herbert Zim, *Seashores: A Guide to Animals and Plants along the Beaches (A Golden Nature Guide)* (New York: Golden Press, 1955).

Photo Credits

All photographs and illustrations in this book are by the authors except the following:

Pages 1–6. Background. © iStockphoto.com/
aimintang.

Pages 8–36. Background. © iStockphoto.com/
AscentXmedia.

Page 18. Waves breaking near Avon, N.C.
Andy Coburn.

Page 18. Spilling breaker. © iStockphoto.com/
DaveAlan.

Page 18. Plunging breaker: © iStockphoto.com/
shannonstent.

Page 18. Surging breaker: © iStockphoto.com/
surfleader.

Pages 27, 29. Rip currents. Miles Hayes.

Pages 30, 35. Background on earth and moon
images. © iStockphoto.com/aopsan.

Page 31. Pier. © iStockphoto.com/myhrcat.

Pages 38–60. Background. © iStockphoto.com/
valio84sl.

Pages 44, 46. Dunes. Orrin Pilkey, William Neal,
Joseph Kelley and Andrew Cooper.

Page 58. Satellite image of an overwash fan.
Google Earth.

Pages 62–74. Background. © iStockphoto.com/
Tolga TEZCAN.

Page 76. Wave ripples. Miles Hayes.

Pages 76–98. Background. Miles Hayes.

Page 77. Current ripples, dune ripples,
ladderback ripples. Miles Hayes.

Page 78. Mysterious feature. Alton Ballance.

Page 79. Near-perfect circle. Orrin Pilkey, William
Neal, Joseph Kelley and Andrew Cooper.

Page 79. Wind pedestal. Larry Fegel.

Page 79. Fossil ripples. Daniel Mayer, Wikmedia
Commons.

Page 84. Air escaping. Bill Neal.

Page 84. Line of bubbles. Bill Neal.

Page 87. Lugworm mounds. Nick Veitch,
Wikimedia Commons.

Page 88. Resting place for a ray. Orrin Pilkey,
William Neal, Joseph Kelley and Andrew
Cooper.

Page 88. Loggerhead turtle tracks. Chris Hart,
USGS.

Page 88. Copperhead. Elizabeth Ray-Schroeder.

Page 92. Cross section of a beach. Miles Hayes.

Pages 100–125. Background. © iStockphoto.com/
EvgenyMinyaev.

Page 112. Brown staining. Rob Greenberg.

Pages 128–58. Background. © iStockphoto.com/
skiserge1.

Page 147. Bitter panicgrass photograph.
USDS-NRCS PLANTS database.

Page 147. American beachgrass photograph.
Royalbroil: Wikimedia Commons.

Page 158. Holes in rock. Lamiot: Wikimedia
Commons.

Pages 160–76. Background. © iStockphoto.com/
temmuz can arsiray.

Page 160. Beach replenishment. © iStockphoto
.com/koosen.

Page 165. Groin at Edisto Beach State Park.
Google Earth.

Page 171. Dead albatross. Chris Jordan,
U.S. Fish and Wildlife Services.

Pages 178–95. Background. © iStockphoto.com/
Warmlight.

Page 179. Diatoms. National Oceanic and
Atmospheric Administration.

Pages 198–204. Background. © iStockphoto.com/
FernandoAH.

Page 200. Comb jelly. National Oceanic and
Atmospheric Administration.

Page 201. Sea turtle. Strobilomyces, Wikimedia
Commons.

About the Authors

Charles Pilkey began his career as a geologist working for an oil company out of Abilene, Texas, an episode he refers to as his Babylonian captivity. By good fortune, he was able to escape by sailing away on a thirty-two-foot sloop, on which he lived for several years, feasting on fish and lobster and seasonally migrating with the birds up and down the East Coast. Like every sailor, he became a lover of the sea for its otherworldly beauty, and a hater of the sea for its savage indifference. Eventually, Charles sold his boat and went back to school to study art. He became a roving sculptor, visiting seven continents and fifty-two countries (including a fourteen-year stint in Japan). The fruits of his travels can be seen in museums and parks and on college campuses in China, Japan, Korea, Italy, Turkey, and the United States.

Orrin Pilkey grew up in eastern Washington State, on the desert side of the mountains. He was very much oriented to the outdoors through fishing, hunting, hiking, canoeing, and even firefighting as a smoke jumper out of Missoula, Montana. Orrin never saw the ocean until he waded into the frigid waters of Puget Sound as a teenager. Like many others who grew up away from the sea, he became fascinated with it and studied oceanography in college. During the first twenty years after graduation, he spent more than a year in total studying the floor of the deep sea. When his parent's house was severely damaged by Hurricane Camille, he realized the sea's true power and decided to trade studying the deep for exploring shorelines. Today, after visiting beaches on all seven continents, Pilkey finds the subject of beaches and how they work more captivating than ever.

Other **Southern Gateways Guides** you might enjoy

How to Read a North Carolina Beach
Bubble Holes, Barking Sands, and Rippled Runnels
ORRIN H. PILKEY, TRACY MONEGAN RICE, AND WILLIAM J. NEAL

A beachcomber's guide to curiosities along the shore

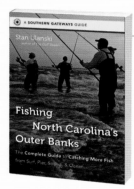

Fishing North Carolina's Outer Banks
The Complete Guide to Catching More Fish from Surf, Pier, Sound, and Ocean
STAN ULANSKI

Improve your fishing techniques (and success) by learning the science of the sea

Life along the Inner Coast
A Naturalist's Guide to the Sounds, Inlets, Rivers, and Intracoastal Waterway from Norfolk to Key West
ROBERT L. LIPPSON AND ALICE JANE LIPPSON
ILLUSTRATIONS BY ALICE JANE LIPPSON

More than 800 species from fresh, brackish, and salty waters

Available at bookstores, by phone at **1-800-848-6224**, or on the web at **www.uncpress.unc.edu**